Corporate Foresight and Innovation Management

This book sheds light on what has come to be known as corporate foresight and its influence on innovation management.

Throughout the book, the contributors examine the practice of corporate foresight and how it may lead to the identification of opportunities for innovation. They also explore the complex processes and conditions that may enable (or impede) the potential of contemporary organizations to capture value from their corporate foresight exercises.

Representing an interesting mix of competing ideas and perspectives, the book offers deep insights into the interactive effects of corporate foresight and its contribution to innovation management.

This book was originally published as a special issue of *Technology Analysis & Strategic Management*.

David Sarpong is a Reader in Strategic Management at Brunel University, London, UK, and Associate Editor for the *Journal of Strategy and Management*. Both his teaching and his research focus on strategy-as-practice, technology and innovation management, and strategic foresight in high-velocity environments.

Dirk Meissner is a Professor at the Research Laboratory for Science and Technology Studies at the Institute for Statistical Studies and Economics of Knowledge, Moscow, Russia. Both his teaching and his research focus on technology and innovation management and policy, with a special focus on foresight and road mapping; science, technology, and innovation policies; funding of research; and priority setting.

Corporate Foresight and Innovation Management

Edited by
David Sarpong and Dirk Meissner

Routledge
Taylor & Francis Group
LONDON AND NEW YORK

First published 2020
by Routledge
2 Park Square, Milton Park, Abingdon, Oxon, OX14 4RN

and by Routledge
52 Vanderbilt Avenue, New York, NY 10017

Routledge is an imprint of the Taylor & Francis Group, an informa business

First issued in paperback 2021

British Library Cataloguing-in-Publication Data
A catalogue record for this book is available from the British Library

ISBN13: 978-0-367-33220-4 (hbk)
ISBN13: 978-1-03-209121-1 (pbk)

Typeset in Myriad Pro
by codeMantra

Publisher's Note
The publisher accepts responsibility for any inconsistencies that may have arisen during the conversion of this book from journal articles to book chapters, namely the inclusion of journal terminology.

Contents

Citation Information

The chapters in this book were originally published in *Technology Analysis and Strategic Management*, volume 30, issue 6 (June 2018). When citing this material, please use the original page numbering for each article, as follows:

Introduction

Special issue on 'corporate foresight and innovation management'
David Sarpong and Dirk Meissner
Technology Analysis & Strategic Management, volume 30, issue 6 (June 2018) pp. 625–632

Chapter 1

Corporate foresight and innovation: the effects of integrative capabilities and organisational learning
Jeewhan Yoon, Youngjun Kim, Nicholas S. Vonortas and Sung Won Han
Technology Analysis & Strategic Management, volume 30, issue 6 (June 2018) pp. 633–645

Chapter 2

Corporate foresight in open laboratories – a translational approach
Albrecht Fritzsche
Technology Analysis & Strategic Management, volume 30, issue 6 (June 2018) pp. 646–657

Chapter 3

Competitive intelligence practices of European firms
Jonathan Calof, Rubén Arcos and Nisha Sewdass
Technology Analysis & Strategic Management, volume 30, issue 6 (June 2018) pp. 658–671

Chapter 4

Fading memories of the future: the dissipation of strategic foresight among middle managers
David Sarpong and Daniël Hartman
Technology Analysis & Strategic Management, volume 30, issue 6 (June 2018) pp. 672–683

Chapter 5

Participation in inter-organisational collaborative open foresight A matter of culture
Melanie Wiener, Regina Gattringer and Franz Strehl
Technology Analysis & Strategic Management, volume 30, issue 6 (June 2018) pp. 684–700

Chapter 6

Potentials of collaborative foresight for SMEs
Yulia Milshina and Konstantin Vishnevskiy
Technology Analysis & Strategic Management, volume 30, issue 6 (June 2018) pp. 701–717

Chapter 7

Overcoming open innovation challenges: a contribution from foresight and foresight networks
Jonathan Calof, Dirk Meissner and Anastasia Razheva
Technology Analysis & Strategic Management, volume 30, issue 6 (June 2018) pp. 718–733

Chapter 8

The role of corporate foresight in exploring new markets – evidence from 3 case studies in the BOP markets
Jakob Højland and René Rohrbeck
Technology Analysis & Strategic Management, volume 30, issue 6 (June 2018) pp. 734–746

Chapter 9

Back to the future of women in technology: insights from understanding the shortage of women in innovation sectors for managing corporate foresight
Vinh Sum Chau and Chanell Quire
Technology Analysis & Strategic Management, volume 30, issue 6 (June 2018) pp. 747–764

Notes on Contributors

Rubén Arcos is a Professor of Communication Sciences and the Deputy Director at the Centre for Intelligence Services and Democratic Systems at Rey Juan Carlos University, Madrid, Spain. He is the Deputy Editor of *The International Journal of Intelligence, Security, and Public Affairs*. In 2013 he founded the SCIP Spain Chapter (Strategic Competitive Intelligence Professionals) and serves as Chapter chair.

Jonathan Calof is a Full Professor of International Business and Strategy at the Telfer School of Management at the University of Ottawa, Canada; an Extraordinary Professor at North-West University, Potchefstroom, South Africa; and a Leading Research Fellow in the Higher School of Economics at the National Research University, Moscow, Russia. He has received the fellow's award from the Strategic and Competitive Intelligence Professionals (SCIP), the lifetime achievement award from Frost and Sullivan, and SCIP's distinguished member award.

Vinh Sum Chau is a Senior Lecturer in Strategy at the Kent Business School at the University of Kent, Canterbury, UK. He is an Associate Editor of the *European Management Review*. He is also a track chair in performance management of the annual British Academy of Management conference.

Albrecht Fritzsche is a Senior Research Fellow at the Institute of Information Systems at Friedrich-Alexander University Erlangen-Nürnberg, Germany, where he teaches innovation strategy and various other topics concerning the organizational and technical conditions of socio-economic change. His current research focusses on the conceptual background of technological change, innovation, collaborative problem solving, and value creation.

Regina Gattringer is an Associate Professor in the Department of Strategic Management at Johannes Kepler University Linz, Austria. Her current research interests focus on open foresight, strategic foresight, open innovation, and R&D networks. She has extensive experience in the implementation of open foresight projects.

Sung Won Han is a Professor in the School of Industrial Management Engineering at Korea University, Seoul, South Korea. He specialises in big data analysis, healthcare, and statistics.

Daniël Hartman completed the UWE MSc International Management in June 2013 at the Bristol Business School at the University of the West of England, UK. His research interests are in organisational learning, strategic foresight, and organising in high-velocity environments.

Jakob Højland is employed at Microsoft as a sales graduate in the Enterprise and Partner Group working with the financial industry. His areas of expertise lie within sales development, new business development, and commercial excellence.

Youngjun Kim is an Associate Dean in the Graduate School of Management of Technology at Korea University, Seoul, South Korea. He specialises in technology management, innovation, and economics.

Dirk Meissner is a Professor at the Research Laboratory for Science and Technology Studies at the Institute for Statistical Studies and Economics of Knowledge, Moscow, Russia. Both his teaching and his research focus on technology and innovation management and policy, with a special focus on foresight and road mapping; science, technology, and innovation policies; funding of research; and priority setting.

Yulia Milshina is a Senior Research Fellow and leading expert in the Foresight centre at the ISSEK at the Higher School of Economics, Moscow, Russia. Her research interests include global technology trends, strategic planning, and foresight.

Chanell Quire was affiliated with the Kent Business School at the University of Kent, Canterbury, UK. She is now a Residential Sales and Leasing Consultant for Better Homes LLC, Dubai.

Anastasia Razheva is a Research Consultant, Project Economist, and Project Manager in Australia. She has gained a great experience in Foresight studies on national and sectoral levels aimed at the long-term forecast of the S&T priorities, development of innovation strategies for state-owned companies, product and technology roadmaps for large companies, and particular sectors (such as shipbuilding and the space industry).

René Rohrbeck is a Professor of Strategy at the Aarhus School of Business and Social Sciences at Aarhus University, Denmark. His areas of expertise are corporate foresight, microfoundations of organisational capabilities, innovation, and technology management.

David Sarpong is a Reader in Strategic Management at Brunel University, London, UK, and an Associate Editor for the *Journal of Strategy and Management*. Both his teaching and his research focus on strategy-as-practice, technology and innovation management, and strategic foresight in high-velocity environments.

Nisha Sewdass is a Professor in the Department of Business Management: Strategy and International Management at the University of South Africa. She is the SCIP Global Advisory Council Lead for Africa and the Founder and Chair of the SCIP (Strategic and Competitive Intelligence Professionals) South African Chapter. She is the recipient of a SCIP Individual Recognition Award and the SCIP Award for Superior Achievement in the Development of Competitive Intelligence Education and the Advancement of the Profession.

Franz Strehl is the Head of the Department of Strategic Management at the Johannes Kepler University Linz, Austria. His current research interests focus on open foresight, strategic foresight, strategic management in R&D networks, organisation analysis, and strategy development and implementation.

Konstantin Vishnevskiy is the Director of the Digital Economics Centre at the ISSEK at the Higher School of Economics, Moscow, Russia. He has long-standing experience in the development of roadmaps, the elaboration of foresight methodology and corporate innovation development programmes, the integration of foresight into government policy, as well as financial and econometric modelling.

Nicholas S. Vonortas is a Professor in the Department of Economics at George Washington University, USA. He specialises in the economics of technological change, science and technology policy, international transfer of technology, and inter-firm cooperation in research and development.

Melanie Wiener is a Postdoctoral Researcher at the Institute for Integrated Quality Design at Johannes Kepler University Linz, Austria. Her research focuses include innovative corporate culture and open foresight.

Jeewhan Yoon is an Assistant Professor in the Department of Management at Cal State LA University, Los Angeles, USA. He specialises in innovation, organisational change, and human resource management.

INTRODUCTION

Corporate foresight and innovation management

David Sarpong and Dirk Meissner

ABSTRACT

This special issue examines corporate foresight and innovation management in contemporary organising. Contributing to a growing body of research on the other-centeredness and interconnectedness of foresight and innovation, the papers in the issue examine the practice of corporate foresight, how it may lead to the identification of opportunities for innovation, and the complex processes and conditions that enable (or impede) the capture of value from corporate foresight. Representing an interesting mix of empirical, conceptual, qualitative, and quantitative methodologies, the papers offer innovative theorising to extend our understanding of the logics of corporate foresight, their interactive effects and contribution to innovation management.

Emphasising anticipation and prescience, corporate foresight is an organising practice involving the creative identification, exploration, and exploitation of opportunities and limits otherwise glossed over by competitors. In fast moving business environments characterised by ambiguity and change, the practice of corporate foresight has become an annual ritual in many forward looking firms (Hodgkinson and Healey 2008; Vecchiato 2015). Following the logic that more of it will lead to companies finding a path that leads to the future, the practice is increasingly conceptualised as a learning process rather than an episodic intervention tool (Bootz 2010; Cunha, Palma, and da Costa 2006). In this regard, a plethora of firms now treat corporate foresight as an everyday practice of combining relevant past, present and future insights into meaningful, future-oriented knowledge (Wright and Cairns 2011; Sarpong and Maclean 2016). This shift has not only altered how firms views and organise in the present to compete in the future. It has led to firms investing in a variety of corporate foresight methodologies ranging from scenario planning, road mapping, competitive intelligence, and business war gaming (Rohrbeck, Battistella, and Huizingh 2015; Daheim and Uerz 2008). Developed and advertised by practitioners, consultants, and researchers as a strategic process of handling the effect and response uncertainty of technology and social drivers of change (Rohrbeck and Gemünden 2011; Vecchiato and Roveda 2010), decision makers are frequently called upon to integrate these exercises into their organisational processes. In response, many organisations now support and conduct scenario planning but continue to struggle to measure their effectiveness or know what kind of future they sometimes want for themselves (Mietzner and Reger 2005).

An expansive view of the corporate foresight 'playing field' now points to inter-organisational collaboration as a strategic means to managing the future. This new turn to participatory foresight projects in clusters and innovation networks seek to mobilise and harness the differential visions of potential competitors and partners to detecting discontinuous changes and interpreting their consequences (Bas and Guillo 2015; Nikolova 2014; Ramos, Mansfield, and Priday 2012). These projects, often sponsored by governments, may have diverse participants, including experts, citizens, and stakeholders whose contributions help the network to expand and build a proprietary 'image of the future' to drive breakthrough innovation (Adegbile, Sarpong, and Meissner 2017; Scuotto et al. 2017). In this regard, this strand of the literature primarily addresses how corporate foresight may stimulate radical and disruptive innovations in innovation clusters (Keller, Markmann, and von der Gracht 2015; van der Duin, Heger, and Schlesinger 2014; Borch, Dingli, and Jørgensen 2013) but have focussed less on extending our understanding of the dynamic interactions between participating firms, the complexities involved in mapping their tasks, connections, architectures, and transaction costs involved in managing such networks.

While it is widely acknowledged among foresight scholars that the link between cause and effect can be elusive, recurrent theme of recent theory on corporate foresight, both on firm and cluster levels, has increasingly found a positive relationship between corporate foresight and innovation (Gershman, Bredikhin, and Vishnevskiy 2016; Ruff 2015; Zeng, Xie, and Tam 2010). Those that do not report direct causal link often identify flexible organisational outcomes such as ambidexterity, peripheral visioning, second-order learning (See for e.g. Paliokaitė and Pačėsa 2015; Rohrbeck and Schwarz 2013; Sarpong and Maclean 2014). These derived outcomes have strategic implications for creation and capture of sustainable value for competitiveness. The emerging consensus, we argue, has come to represent something of a conundrum. Those that have attempted to examine the causal link between corporate foresight and innovation frequently struggle to show exactly when and how foresight exercises may lead to innovation while others simply neglect the complexities involved in mapping the connections and linkages between foresight practices and innovation (Sarpong, Maclean, and Davies 2013; Rohrbeck and Gemünden 2011; Miles et al. 2017).

We accept that strategic foresight as a discipline is still in its pre-paradigmatic stage but the conceptual landscape has become extremely fragmented with the potential to obfuscate theoretical leaps aimed at improving the practice in organising. In particular, theoretical and empirical specifications on how strategic foresight may lead to the identification of opportunities for innovation remain sparse (Rohrbeck and Gemünden 2011). As noted in the original call for papers, the purpose of this special issue, therefore, is to contribute to the progress being made by the broader foresight literature and augment recent conceptual and empirical developments that examine the limits and potentialities of corporate foresight as a bridge to innovation and innovation management in general.

Overview of articles in this special issue

The papers contained in this special issue do not only reflect the current state of the corporate foresight and innovation literature. As a group, these articles demonstrate the extent to which foresight scholars are contributing to new ways to stimulating and improving organisational foresightfulness. We delineate the nine papers in this special issue along three themes reflecting new thinking about corporate foresight and innovation management: Leveraging innovation from corporate foresight at the firm level; inter-organisational participative foresight and the identification of opportunities for innovation; and inclusivity in corporate foresight and innovation. Table 1 is a summary of the papers contained in the issue.

Leveraging innovation from corporate foresight at the firm level

Evaluating one of the fundamental assumption driving corporate foresight practices, Yoon et al. (2018) empirically examined the influence of corporate foresight on firm innovativeness of a variety of Korean manufacturing firms. Drawing on the resource-based view and dynamic capabilities, they developed a model and posited that the effect of corporate foresight on innovativeness is mediated by organisational learning, while the relationship between corporate foresight and organisational learning is moderated by integrative capabilities. The study showed that the indirect effect

Table 1. Overview of contributions.

Authors	Research questions	Theoretical lens	Empirical setting (data and methods)	Key findings
Jeewhan Yoon, Youngjun, Kim, Nicholas S. Vonortas and Sung Won Han	What is the effect of corporate foresight on innovativeness?	Foresight, Resource Based View (RBV), and dynamic capabilities	South Korea based manufacturing firms; Quantitative survey	Built on a model that shows that the influence of corporate foresight on innovativeness is mediated by organisational learning. The indirect effect they found is conditioned by organizational learning and integrative capabilities of the firm.
Albrect Fritzshe	How do the translation practices of open laboratories contribute to corporate foresight?	Contemporary turn to translation in cultural theory	Open laboratory in a major European city; Processed-based qualitative	Open-laboratories bring contrasting under-standing of innovation to life by opening up opportunities for corporate foresight exercises to understand different stakeholders' attributions of meaning inside and outside the firm.
Jonathan Calof, Rubén Arcos and Nisha Sewdass	What is the current state of competitive intelligence (CI) in European firms?	Competitive intelligence (CI) and foresight practices	Competitive intelligence (CI) practices in European firms; Comparative survey	CI in European firms has evolved beyond competitor analysis to R&D and new product innovation. The study also found similarities between foresight and CI in terms of objective setting and the use of scenarios and environmental analysis in developing images of the future.
David Sarpong and Daniël Hartman	How do organizing process and routines lead to the dissipation of strategic foresight among middle managers?	Strategic foresight as a distributed capability	Single case study of a European retail firm; qualitative interviews	In examining how organizing processes and routines may dissipate foresight in practice, the study highlights how the legitimation of rhetoric, instrumental rationality, suppression of creative freedom, and the formulations of grand solutions in search of problems may cumulatively lead to the dissipation of strategic foresight among middle managers.
Melanie Weiner, Regina Gattringer and Franz Strehl	Which types of corporate cultures are likely to foster or inhibit a firm's openness for collaborative open foresight?	Open-innovation and competing values framework	Two Austrian firms with experience in collaborative foresight projects; Interviews	Organizations characterised by clan or adhocracy cultures tend to foster openness. At the extreme end of the continuum, market and hierarchy cultures inhibits openness for collaborative open foresight participation.
Yulia Milshina and Konstantin Vishneevskiy	How can collaborative foresight help SMEs to become competitive?	Strategic foresight and innovation clusters	SME embedded in a new Russian medical technologies cluster; Road-mapping	Collaborative foresight projects provide SMEs the opportunity to overcome their resource constraints to identify potential technology chains that can be translated into innovative priorities and indicators to build plausible futures for SMEs.
Jonathan Calof, Dirk Meissner, D and Anastasia Razheva	What can foresight influence successful implementation of open innovation?	Foresight networks and knowledge integration	Conceptual review of literature on foresight and open innovation	Applied at each stage of the innovation value chain, foresight and foresight networks can help in the identification of challenges in the implementation of open innovation systems.

(*Continued*)

Table 1. Continued.

Authors	Research questions	Theoretical lens	Empirical setting (data and methods)	Key findings
Jacob Højland and René Rohrbeck	To what extent can business-development activities in BoP markets be classified as corporate foresight?	Foresight and innovation at the base of the pyramid (BoP)	Case studies-three Danish firms; company reports and documents, and interviews	The study found that both cognitive and experimental search elements of strategic foresight, such as probing and prospecting are important in exploring for opportunities and limits in BoP markets.
Vinh Sum Chau and Chanell Quire	What can foresight tell us about the shortage of women in innovation, and science, technology, and engineering and mathematics (STEM) sectors?	Gender in foresight, science, and technology	UK STEM sectors; Interviews with female scientists, and questionnaire	This study examined the shortage of women in the STEM sector, and how their active inclusion can improve corporate foresight in the technology sector as a whole. Findings from the study suggest that greater inclusion of women in foresight exercises in the STEM sectors will broaden the futures of the predominantly male-dominated sector. Three case-rich narratives are presented as a summary on the future of women in technology.

of corporate foresight on innovativeness through organisational learning is conditioned by the level of integrative capabilities possessed by a firm.

In taping into the crowds wisdom to strengthen corporate foresight exercises, Fritzsche (2018) draws on the recent turn to translation in cultural theory to examined the practices of 'open laboratories' and their value added potential to enhancing corporate foresight. Focussing on the semantic differences in the differential ways different groups of people express meaning, the study conducted in an open laboratory located in the centre of a major European city unpacks the differential perspectives of firms and their customers on how they can improve their business processes. Findings from the study suggest that translation in open laboratories can enable companies to gain insight into their own capabilities, environment, and value networks. Such knowledge, the paper argues, can be fed into corporate foresight exercises to enhance engagement and on-going (re)negotiation of a firm's relation with stakeholders in commercial interactions.

Shedding light on how competitive intelligence has become part of everyday organising, Calof, Arcos, and Sewdass (2018) presents their very recent results of a survey of competitive intelligence (CI) practices in European firms. In comparing the recent results to a 2006 global study and a 2006 European CI study, they found that the breadth of applications for CI has grown well beyond competitors to include customer related intelligence, technology, market, etc. The go further to posit that innovation is driving much intelligence activity, in particular research and development (R&D) and new product development decisions, and CI appears to be much more formalised in European firms than it was in 2006. In particular, the study also found similarities between corporate foresight and CI in terms of objectives (development and maintenance of competitive advantage, help with decision making) and analytical techniques with scenarios being among the more frequently used analytical techniques along with STEEP and other environmental analysis in both corporate foresight studies and the CI study.

Also focussing on the meso level, Sarpong and Hartman (2018) examined organising practices and routines that potentially combine to impede strategic foresight among middle managers. Building on an explorative case study of a European sportswear retail company that has undergone ownership and structural changes on how work was organised. Extending our understanding of contextual antecedents that could lead to the dissipation of strategic foresight among middle managers in organising, the study identifies four organising processes and their associated routines (rhetoric of legitimation, instrumental rationality, suppression of creative freedom, and the formulations of solutions in search of problems) which typify the reported patterns of foresight dissipation among middle managers.

Inter-organisational participative foresight

The paper by Wiener, Gattringer, and Strehl (2018) explores the influence of organisational culture on firms' ability to benefit from inter-organisational collaborative open foresight projects. Based on two in-depth case studies of organisations that have engaged in one of such participative projects, the study draws on the Competing Values Framework for (organisational) culture as an organising framework to examine how the collective mental programming of the firm could impact its 'openness' and potential to benefit from participative foresight projects. Findings from the study suggest that firms characterised as having a clan or adhocracy cultures are likely to foster openness, where as those characterised as having a market or hierarchy cultures tend to stifle openness for open foresight participation, and in turn struggle to exploit the value generated by the network to remain competitive.

The paper by Milshina and Vishnevskiy (2018) reporting on a Russian foresight project that employed road mapping as a methodological tool to examine the potential of collaborative foresight project in cluster made up of SMEs pioneering advanced medical technologies. Findings from the study suggest the project did not only provide a context for interaction for participating firms, but also enabled them to form self-organising technology chains and build linkages with research universities. In making the case for SMEs to engage in collaborative foresight projects to enhance their innovation capacity, some of the major innovative achievements of the cluster are delineated. The paper concludes that cluster driven collaborative foresight projects may serve as quintessential laboratories for SMEs, particularly, those with limited R&D budgets to pull their scarce resources together to explore emerging technologies with far reaching applications and potential.

The conceptual paper by Calof, Meissner, and Razheva (2018) combines concepts from foresight and foresight networks to explore how corporate foresight in practice could be leveraged to drive open innovation. Organised around key open innovation questions such as those related to technology selection, identifying future customer needs and scanning for disruptions, the paper provides some useful insight into how foresight can potentially be exploited to address some of the inherent challenges blighting open innovation. From this perspective, they argue that the weaknesses in open innovation lies in the strength of foresight as a tool for obtaining access to appropriate external experts and their knowledge; and making sense of the mass information that can emerge through a more open process. They conclude highlighting how foresight networks could point open innovation to new ideas in innovative collaboration forms.

Inclusivity in corporate foresight and innovation

Drawing on interviews with five senior personnel and a survey of 365 females working in a variety of firms in the science, technology, engineering and mathematics (STEM) sectors, Chau and Quire (2018) investigated why there is a shortage of women in this sector and offers insights for managing corporate foresight. They found that their incisive strategic minds which could spark imaginations and audacious visions of the future are often overlooked and marginalised because they tend to inhabit lower levels of the organisational hierarchy. They suggest that their tolerance for outliner views and opinions can be harnessed to shape alternative scenarios and groupthink in organising. They conclude that greater inclusion of women in corporate foresight exercises can help shape the futures and destinies of yet-to-be-realized innovations.

Drawing on evidence from three cases studies in the Base of the Pyramid (BOP) markets, Højland and Rohrbeck (2018) explored the strategic role of corporate foresight in exploring new markets. They examined the extent to which successful business-development activities in uncertain environments can be classified as corporate foresight (CF) and the extent to which they have been intentional and systematic. Emphasising the need for corporate foresight to narrow down on experimental and cognitive search mechanisms to probing uncertain environments, findings from the their study suggest that the probing (experimental search) phase is particularly important in unknown and uncertain environments but that perceiving and prospecting (cognitive search) activities are necessary to find distant opportunities. They conclude by positing that successful business-development activities rely on multiple iterations between perceiving, prospecting and probing.

Conclusions and potential future research avenues

Taken together, the nine papers contained in this special issue do not only shed light on some of the salient theoretical and empirical foundations underlying strategic foresight and its relationship to innovation. They also indicate the current trends in strategic foresight research. The following are some directions for future research in this domain:

(1) How can we assess and measure the return-on-foresight exercises and develop foresight capabilities in sustainable ways?
(2) What are the institutional antecedents that enable (or impede) value capture from participative corporate projects?
(3) What is the future of participative corporate projects?
(4) How can SMEs leverage corporate foresight exercises to determine technology futures?

(5) When do corporate foresight exercises lead to anticipatory intelligence needed to decide on competing or alternative technological pathways?

We encourage scholars to explore these research avenues drawing on the wide variety of theoretical, conceptual, and methodological approaches in their tool kits. We are of the firm belief that the proposed directions for future research, and the papers contained in this volume will spur other foresight researchers and practitioners to continue with their efforts to bridging the epistemic gaps between corporate foresight and innovation.

Acknowledgement

Dirk Meissner's contribution to the article was prepared within the framework of the Basic Research Programme at the National Research University Higher School of Economics (HSE) and supported within the framework of the subsidy granted to the HSE by the Government of the Russian Federation for the implementation of the Global Competitiveness Programme.

Disclosure statement

No potential conflict of interest was reported by the authors.

ORCID

Dirk Meissner http://orcid.org/0000-0003-4573-0376

References

Adegbile, A., D. Sarpong, and D. Meissner. 2017. "Strategic Foresight for Innovation Management: A Review and Research Agenda." *International Journal of Innovation and Technology Management* 14 (04): 1750019.

Bas, E., and M. Guillo. 2015. "Participatory Foresight for Social Innovation. FLUX-3D Method (Forward Looking User Experience), a Tool for Evaluating Innovations." *Technological Forecasting and Social Change* 101: 275–290.

Bootz, J. P. 2010. "Strategic Foresight and Organizational Learning: A Survey and Critical Analysis." *Technological Forecasting and Social Change* 77 (9): 1588–1594.

Borch, K., S. M. Dingli, and M. S. Jørgensen. 2013. *Participation and Interaction in Foresight: Dialogue, Dissemination and Visions*. Cheltenham: Edward Elgar Publishing.

Calof, J., R. Arcos, and N. Sewdass. 2018. "Competitive Intelligence Practices of European Firms." *Technology Analysis & Strategic Management* 30 (6): 658–671. doi:10.1080/09537325.2017.1337890.

Calof, J., D. Meissner, and A. Razheva. 2018. "Overcoming Open Innovation Challenges: A Contribution from Foresight and Foresight Networks." *Technology Analysis & Strategic Management* 30 (6): 718–733. doi:10.1080/09537325.2017.1351609.

Chau, V. S., and C. Quire. 2018. "Back to the Future of Women in Technology: Insights from Understanding the Shortage of Women in Innovation Sectors for Managing Corporate Foresight." *Technology Analysis & Strategic Management* 30 (6): 747–764. doi:10.1080/09537325.2017.1376046.

Cunha, M. P., P. Palma, and N. G. da Costa. 2006. "Fear of Foresight: Knowledge and Ignorance in Organisational Foresight." *Futures* 38 (8): 942–955.

Daheim, C., and G. Uerz. 2008. "Corporate Foresight in Europe: From Trend Based Logics to Open Foresight." *Technology Analysis & Strategic Management* 20 (3): 321–336.

Fritzsche, A. 2018. "Corporate Foresight in Open Laboratories–A Translational Approach." *Technology Analysis & Strategic Management* 30 (6): 646–657. doi:10.1080/09537325.2017.1380180.

Gershman, M., S. Bredikhin, and K. Vishnevskiy. 2016. "The Role of Corporate Foresight and Technology Roadmapping in Companies' Innovation Development: The Case of Russian State- Owned Enterprises." *Technological Forecasting and Social Change* 110: 187–195.

Hodgkinson, G. P., and M. P. Healey. 2008. "Toward a (Pragmatic) Science of Strategic Intervention: Design Propositions for Scenario Planning." *Organization Studies* 29 (3): 435–457.

Højland, J., and R. Rohrbeck. 2018. "The Role of Corporate Foresight in Exploring New Markets– Evidence from 3 Case Studies in the BOP Markets." *Technology Analysis & Strategic Management* 30 (6): 734–746. doi:10.1080/09537325.2017.1337887.

Keller, J., C. Markmann, and Heiko A. von der Gracht. 2015. "Foresight Support Systems to Facilitate Regional Innovations: A Conceptualization Case for a German Logistics Cluster." *Technological Forecasting and Social Change* 97: 15–28.

Mietzner, D., and G. Reger. 2005. "Advantages and Disadvantages of Scenario Approaches for Strategic Foresight." *International Journal of Technology Intelligence and Planning* 1 (2): 220–239.

Miles, I., D. Meissner, N. S. Vonortas, and E. Carayannis. 2017. "Technology Foresight in Transition." *Technological Forecasting & Social Change* 119: 211–218.

Milshina, Y., and K. Vishnevskiy. 2018. "Potentials of Collaborative Foresight for SMEs." *Technology Analysis & Strategic Management* 30 (6): 701–717. doi:10.1080/09537325.2017.1406906.

Nikolova, B. 2014. "The Rise and Promise of Participatory Foresight." *European Journal of Futures Research* 2 (33): 1–9.

Paliokaitė, A., and N. Pačėsa. 2015. "The Relationship between Organisational Foresight and Organisational Ambidexterity." *Technological Forecasting and Social Change* 101: 165–181.

Ramos, J., T. Mansfield, and G. Priday. 2012. "Foresight in a Network Era: Peer-Producing Alternative Futures." *Journal of Futures Studies* 17 (1): 71–90.

Rohrbeck, R., C. Battistella, and E. Huizingh. 2015. "Corporate Foresight: An Emerging Field with a Rich Tradition." *Technological Forecasting and Social Change* 101: 1–9.

Rohrbeck, R., and H. Gemünden. 2011. "Corporate Foresight: Its Three Roles in Enhancing the Innovation Capacity of a Firm." *Technological Forecasting and Social Change* 78 (2): 231–243.

Rohrbeck, R., and J. O. Schwarz. 2013. "The Value Contribution of Strategic Foresight: Insights from an Empirical Study of Large European Companies." *Technological Forecasting and Social Change* 80 (8): 1593–1606.

Ruff, F. 2015. "The Advanced Role of Corporate Foresight in Innovation and Strategic Management—Reflections on Practical Experiences from the Automotive Industry." *Technological Forecasting and Social Change* 101: 37–48.

Sarpong, D., and D. Hartman. 2018. "Fading Memories of the Future: The Dissipation of Strategic Foresight Among Middle Managers." *Technology Analysis & Strategic Management* 30 (6): 672–683. doi:10.1080/09537325.2017.1376736.

Sarpong, D., and M. Maclean. 2014. "Unpacking Strategic Foresight: A Practice Approach." *Scandinavian Journal of Management* 30 (1): 16–26.

Sarpong, D., and M. Maclean. 2016. "Cultivating Strategic Foresight in Practise: A Relational Perspective." *Journal of Business Research,* 69 (8): 2812–2820.

Sarpong, D., M. Maclean, and C. Davies. 2013. "A Matter of Foresight: How Practices Enable (or Impede) Organizational Foresightfulness." *European Management Journal* 31 (6): 613–625.

Scuotto, V., M. Del Giudice, S. Bresciani, and D. Meissner. 2017. "Knowledge-Driven Preferences in Informal Inbound Open Innovation Modes. An Explorative View on Small to Medium Enterprises." *Journal of Knowledge Management* 21 (3): 640–655.

van der Duin, P., T. Heger, and M. D. Schlesinger. 2014. "Toward Networked Foresight? Exploring the use of Futures Research in Innovation Networks." *Futures* 59: 62–78.

Vecchiato, R. 2015. "Creating Value Through Foresight: First Mover Advantages and Strategic Agility." *Technological Forecasting and Social Change* 101: 25–36.

Vecchiato, R., and C. Roveda. 2010. "Strategic Foresight in Corporate Organizations: Handling the Effect and Response Uncertainty of Technology and Social Drivers." *Technological Forecasting and Social Change* 77: 1527–1539.

Wiener, M., R. Gattringer, and F. Strehl. 2018. "Participation in Inter-Organisational Collaborative Open Foresight: A Matter of Culture." *Technology Analysis & Strategic Management* 30 (6): 684–700. doi:10.1080/09537325.2017.1376045.

Wright, G., and G. Cairns. 2011. *Scenario Thinking: Practical Approaches to the Future.* Cheltenham: Palgrave Macmillan.

Yoon, J., Y. Kim, N. S. Vonortas, and S. W. Han. 2018. "Corporate Foresight and Innovation: The Effects of Integrative Capabilities and Organisational Learning." *Technology Analysis & Strategic Management* 30 (6): 633–645. doi:10.1080/09537325.2017.1395407.

Zeng, S. X., X. M. Xie, and C. M. Tam. 2010. "Relationship Between Cooperation Networks and Innovation Performance of SMEs." *Technovation* 30 (3): 181–194.

Corporate foresight and innovation: the effects of integrative capabilities and organisational learning

Jeewhan Yoon*, Youngjun Kim*, Nicholas S. Vonortas and Sung Won Han

ABSTRACT

Despite the importance of corporate foresight for innovation management, scholars have yet to identify the organisational processes through which corporate foresight influences a company's innovativeness. Drawing on the resource-based view and dynamic capabilities theory, we developed a model and posited that the effect of corporate foresight on innovativeness is mediated by organisational learning, while the relationship between corporate foresight and organisational learning is moderated by integrative capabilities. We conducted an empirical study in the manufacturing industry and found support for the study model, in which the indirect effect of corporate foresight on innovativeness through organisational learning was conditional on the level of integrative capabilities. This study has theoretical and practical implications for a more nuanced understanding of corporate foresight and its effect on innovation management.

Introduction

In order to survive and succeed in today's complex and volatile business environment, companies increasingly rely on corporate foresight (Rohrbeck and Gemünden 2011). Corporate foresight is an organisational resource that includes any cultural or structural element that enables the company to detect change early, identify new promising business fields, analyze the consequences to the company, and formulate effective responses to sustain innovation (Sarpong, Maclean, and Alexander 2013). Many companies, accordingly, develop corporate foresight practices to lay the foundation for a future competitive advantage and manage the risk of technology change (Vecchiato and Roveda 2010).

The corporate foresight literature has generally found a positive relationship between corporate foresight and organisational outcomes such as innovation management and value creation, which can promote new product development and technology advancement (Vishnevskiy, Karasev, and Meissner 2015; Ramírez, Roodhart, and Manders 2011; Rohrbeck 2012). However, scholars who have attempted to examine this positive link have frequently struggled to show exactly how and when corporate foresight can lead to innovation (Rohrbeck and Gemünden 2011). Additionally, the contextual conditions of corporate foresight in stimulating innovation are often unknown due to the theoretical and methodological complexities involved in mapping intricate routes in the

*These two authors contributed equally to this manuscript.

connection, structure, and influence of corporate foresight on innovation management (Georghiou and Harper 2011).

Although the positive relationship between corporate foresight and innovation seems to be intuitive, how corporate foresight can contribute to the innovativeness of the organisation still depends on whether organisations are engaged in organisational learning, such as discussing alternatives, experimenting with new ideas, and solving problems to come up with the best possible way to innovate products, services, and work processes. This study proposes organisational learning as a mediator in the relationship between corporate foresight and innovativeness. The resource-based view (RBV) and organisational learning literatures suggest that organisational learning may be a core underlying mechanism that explains how corporate foresight can be leveraged when an organisation chooses among competing and alternative technological pathways to pursue its future (Sarpong and Maclean 2012).

In addition, Rohrbeck and Gemünden (2011) have argued that more research is needed to identify any boundary conditions that may influence the manifestation of innovation. Analyzing the moderating effect of organisational factors will help depict the condition under which corporate foresight is most likely to result in organisational learning, and thus innovativeness. In this study, we respond to Rohrbeck and Gemünden's (2011) call by investigating a contextual factor that is likely to influence the extent to which organisations with corporate foresight choose to enact organisational learning. Drawing on dynamic capabilities theory, we refer to organisations' capability to integrate knowledge from multiple, distributed sources within its boundaries and beyond as integrative capabilities. An organisation's integrative capabilities include competence in utilising knowledge and skills from multiple sources, and these capabilities can strengthen the relationship between corporate foresight and organisational learning (Nadler, Tushman, and Nadler 1997). This study examines whether the relationship between corporate foresight and organisational learning is strengthened by high integrative capabilities.

The purpose of this paper, therefore, is to identify and examine specific processes through which organisations' corporate foresight can influence their innovativeness and to determine the relevance and contribution of corporate foresight to innovation management. Specifically, drawing on RBV and dynamic capabilities theory, this paper aims to identify organisational antecedents such as organisational learning and integrative capabilities that enable the creative emergence of innovation from corporate foresight exercises. By examining a possible mediator and moderator of the relationship between corporate foresight and innovativeness, this study contributes to the understanding of the role of corporate foresight for innovation management.

Theoretical background and hypothesis development

The impact of corporate foresight on innovativeness

Studies have emphasised the importance of corporate foresight for innovation (e.g. Ruff 2015), with most studies suggesting a positive relationship between corporate foresight and innovation (Ramírez, Roodhart, and Manders 2011). For example, recent studies have focused on how corporate foresight exercises can be used to better predict the promise of technologies and various business opportunities (Battistella 2014; Gershman, Bredikhin, and Vishnevskiy 2016; Khripunova et al. 2014). Yam and his colleagues (2004) developed the innovation audit framework to investigate the determinant factors of innovation outcomes of Chinese firms. Their results showed that corporate foresight and technology capabilities had an effect on innovation.

Through an empirical investigation, Rohrbeck and Gemünden (2011) identified specific reasons that corporate foresight can enhance the innovativeness of an organisation. First, by taking an *initiator* role, corporate foresight supports innovation because it identifies emerging technologies and new customer needs through various methods such as technology scouting and market surveys. Corporate foresight might also support innovation through the monitoring of competitors to analyze how they respond to changes in technology and customer needs. In addition, corporate foresight enables a company to identify external expert networks that can be used later to acquire the necessary technologies (Rohrbeck 2010b). Second, by taking a *strategist* role, corporate foresight supports innovation by identifying gaps between the current status and goal status of the company. To address the gap, corporate foresight focuses on the internal renewal of work processes and the repositioning of efforts of the company in the market. Finally, by taking an *opponent* role, corporate foresight uses the insights from environmental scanning to challenge the company's ongoing research and development projects. Corporate foresight can be used to ensure that innovation initiatives are continuously benchmarked against emerging technologies and confronted with current and future customer needs. Thus, we posit the following hypothesis:

Hypothesis 1: Corporate foresight is positively related to innovativeness.

The relationship between corporate foresight and organisational learning

RBV-based research has shown that companies use strategic resources such as corporate foresight to obtain an edge over their competitors. Specifically, RBV suggests that maintaining competitive advantage requires resources that are rare, inimitable, and sustainable (Collis and Montgomery 1995), along with the ability to employ those resources (Barney 1986). Each firm has its own resource endowments (Barney 1991), but additional capabilities may be needed for it to compete at different points in time and in different markets (Hitt et al. 1999). Organisations thus need to acquire new knowledge and learn new technical and managerial capabilities on an ongoing basis (Mothe and Quelin 1998).

The key motives for engaging in organisational learning are utilising existing organisational resources such as corporate foresight and accumulated technologies to obtain competitive advantage. Through organisational learning, organisations utilise existing knowledge and access new skills from internal and external environments (see Khanna, Gulati, and Nohria 1998) so as to enhance their own competencies in new product development (Rothaermel 2001). Studies have shown that organisational learning infuses an organisation with new ideas (Hurley and Hult 1998), enhances the capacity to understand new ideas (Damanpour 1991), enhances creativity and the ability to notice novel opportunities (Van de Ven and Angle 1989), nurtures and encourages innovative ideas (Waldman and Bass 1991), and signals to employees that they are valued, which encourages them to care about innovation for the good of the organisation (Waldman and Bass 1991).

Additionally, organisational learning illustrates how organisations can develop their corporate foresight to acquire new capabilities and use those capabilities to create a competitive advantage (Lichtenthaler 2009). Thorpe and his colleagues (2005) argued that organisational learning enables a company to develop and exploit innovation networks with other entities, even its competitors. Recently, the term *open innovation* has been used to express the learning of a company to use its corporate foresight and collaborate with others in order to sense any changes and acquire any needed capabilities (Rohrbeck and Gemünden 2011). Based on the findings and theoretical arguments outlined, we posit the following hypothesis:

Hypothesis 2: Corporate foresight is positively related to organizational learning.

The mediating role of organisational learning in the relationship between corporate foresight and innovativeness

Scholars have acknowledged that corporate foresight and any other organisational resources, by themselves, might not be the main route for innovation. That is, resources and capabilities can be a source of business performance if they are used and exploited through organisational and environmental processes (Russo and Fouts 1997).

The literatures of innovation and organisational learning also suggest that many organisations foster organisational processes such as organisational learning so as to lead corporate foresight to innovation (Shrivastava and Grant 1985). Specifically, studies have shown that companies need to engage with learning behaviours such as environmental scanning and experimenting in order to obtain up-to-date knowledge about the direction and magnitude of emerging external change (May, Stewart, and Sweo 2000). Organisational learning is important because innovation is often characterised by how organisations win through short periods of radical change through learning, which is the activity to explore new business opportunities and to adapt to new markets in times of discontinuous change (Tushman, Virany, and Romanelli 1985).

We consider organisational learning a plausible intermediate process through which corporate foresight can improve innovation. Studies have highlighted how corporate foresight affects outcomes such as visioning and ambidexterity, which promote technology advancements and new product development in unstable environments (Anderson et al. 2004; Von der Gracht, Vennemann, and Darkow 2010). It has been presumed – but not tested – that organisational learning accounts for the effects of corporate foresight on innovation. By contrast, when organisational learning is the outcome of a study, researchers have rarely included other outcomes, assuming the performance benefit of organisational learning (Huber 1991). Addressing these shortcomings of the organisational learning and corporate foresight literature, we propose that organisational learning is an important venue through which corporate foresight can foster innovation. When organisations have corporate foresight, with its capabilities and resources, they will feel comfortable engaging in learning, and organisational learning to expand and fully utilise its knowledge may boost innovation because such learning leads to the identification of novel approaches and procedures to apply to organisational goals (Griffith and Sawyer 2010).

Hypothesis 3: Organizational learning mediates the relationship between corporate foresight and innovativeness.

Integrative capabilities as a moderator of the corporate foresight–organisational learning behaviours relationship

Developing a new product or introducing process innovations is challenging – requiring the ability to integrate the various functions, activities, and information flows within and between organisations (Lorenzoni and Lipparini 1999). An organisation's integrative capabilities, a competence in integrating knowledge from multiple, distributed sources within organisational boundaries and beyond (Kogut and Zander 1992), involve the capacity to use existing knowledge from internal sources, absorb knowledge from external sources, and blend the different technical competencies required for innovation.

Ambrosini and Bowman (2009) extended RBV and argued that organisations need to refresh existing resources into valuable and rare capabilities in a changing environment. Organisations need to combine and develop new resources consistently to maintain their competitive advantage (Eisenhardt and Martin 2000). The theory of dynamic capabilities also suggests that companies faced with disruptions in the environment (Arnold 2003) need to consolidate and apply their various strategic resources (e.g. their corporate foresight exercises) to regain a competitive advantage (Teece, Pisano, and Shuen 1997). Thus, organisations with integrative capabilities tend to be able to identify and combine needed capabilities and to facilitate their learning or development over time (Helfat and Peteraf 2003).

Cohen and Levinthal (1990) suggested that a company's ability to detect the value of external information, integrate it, and apply it to learning and development is critical to its innovative performance. The problem faced by many companies is that even if a new technology is successfully developed, this technology is usually just one piece of knowledge that needs to be complemented with other developments (i.e. subsystems, process innovations) so as to be embodied in a complex product that delivers the functionality required for commercial application. The effectiveness of a company's attempt to appropriate and further develop knowledge spillovers depends upon sufficient levels of integrative capabilities.

Specifically, relying on RBV and dynamic capabilities theory, we propose that the extent of integrative capabilities possessed by an organisation has a positive association with the relationship between corporate foresight and organisational learning. The ability of an organisation to combine knowledge and consolidate diverse competencies is likely to play a contextual role in its engagement in learning behaviour (Edmondson 1999). As organisations need to combine and develop new resources consistently to maintain their competitive advantage (Eisenhardt and Martin 2000), those with more integrative capabilities have the resources to solve problems and deal with difficult situations. Ruff (2015) argued that corporate foresight practices need a flexible, multidisciplinary, and open environment to enhance strategy and innovation. The relationship between corporate foresight and organisational learning should be stronger when an organisation's integrative capabilities are high than when they are low.

> *Hypothesis 4:* The positive relationship between corporate foresight and organizational learning is moderated by integrative capabilities, so that it is stronger for higher levels than for lower levels of integrative capabilities.

The study model is illustrated in Figure 1, showing that the indirect effect of corporate foresight on innovativeness through organisational learning depends on the extent of integrative capabilities.

Method

Sample and procedure

Our study sample consisted of South Korean companies in the manufacturing industry, which transform raw materials into intermediary or finished goods by using various tools such as biological processing, chemicals, labour, and machines. Those responsible for the human resources (HR) function of the participating companies acknowledged the value of strategic initiatives and organisational work processes and identified employees who provided a good representation of the overall workforce. After explaining that their information would be kept confidential, we invited these employees to take part in the survey, using time at work, with a deadline of 2 weeks. The majority of respondents (78.8%) were men; respondents' average age was 41.61 years (standard deviation [SD] = 11.22), and their tenure at the organisation averaged 7.52 years (SD = 5.65). Most respondents worked as machine operators (41.7%) or assembly line coordinators (30.7%); other occupations included customer service representatives (13.0%), administrators (8.6%), and managers or senior officials (6%).

This study considered organisation-level variables. Organisation-level variables were derived by aggregating responses of individual organisational members responding to the survey questionnaire asking their perception of organisation-level measures (e.g. corporate foresight and organisational learning); the data of individuals who belonged to the same organisation were statistically aggregated to represent the organisation-level measures of the study (Christie 1990). Thus, the aggregated

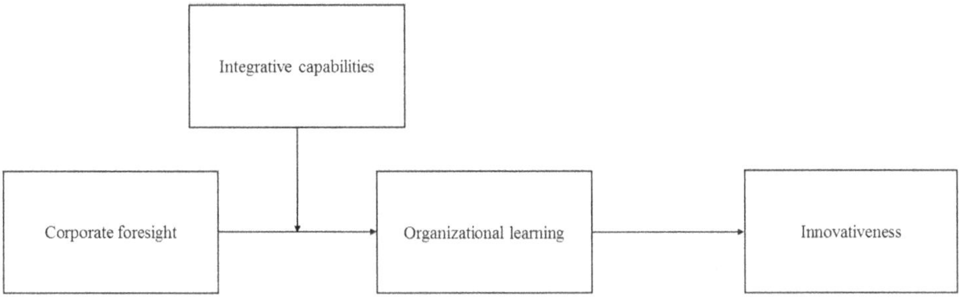

Figure 1. Summary model of the study.

variable represented not just the sum of individual perceptions but an organisation-level shared perception of the organisation (Firebaugh 1978). A total of 346 companies were sampled, and 243 participated. After deletion of missing values, the final sample consisted of 212 companies.

We also used several methods to control for the influence of common method bias. Following the recommendations of Podsakoff et al. (2003), we used established scales, explained procedures to our participants, and guaranteed participant anonymity. We also used filter items and different instructions to create a psychological separation between the sets of variables.

Measures

Corporate foresight

To measure corporate foresight, we used Rohrbeck and Gemünden's (2011) 11-item instrument that considers the initiator, strategist, and opponent roles of corporate foresight. Internal consistency was 0.82. Responses for these questions, as well as for all others on the survey, used a 7-point Likert scale that ranged from 1 (strongly disagree) to 7 (strongly agree).

Organisational learning

Organisational learning was measured with a 7-item instrument developed by Burke (1989) and adopted by Hurley and Hult (1998). Their measure has been used in the past as an antecedent to innovativeness and the capacity to innovate and has been found to be reliable and valid on the basis of commonly accepted procedures (Hurley and Hult 1998). Internal consistency was 0.88.

Integrative capabilities

We measured organisations' integrative capabilities with a 4-item instrument adopted by Helfat and Raubitschek (2000) and Spanos, Vonortas, and Voudouris (2015). This measure assesses the organisation's ability to integrate cross-functional activities within and across organisational boundaries and develop strategies that integrate market needs into their offerings. Internal consistency was 0.92.

Innovativeness

Two weeks after measuring the other variables, we measured organisational innovativeness using the 5-item scale adapted by Hurley and Hult (1998) to measure innovativeness of the company or attitudes towards innovation. Internal consistency was 0.84.

Control variables

We controlled for organisational size and age. As others have done (e.g. Hmieleski and Baron 2009), we estimated size based on the number of people in the organisation, using data acquired from HR directors. The mean organisational size for the sample was 14.50 (SD = 4.25), and the mean organisational age was 17.43 years (SD = 5.99).

Analyses

We used hierarchical regression with mean-centered predictor variables (Aiken, West, and Reno 1991) to test Hypotheses 1 and 2 and used the approach described by Preacher, Rucker, and Hayes (2007) for conditional indirect effects to test Hypotheses 3 and 4. Also, for Hypothesis 3, we followed the steps outlined by Baron and Kenny (1986) for mediation, establishing that (1) the independent variable correlates with the dependent variable, (2) the independent variable correlates with the mediator variable, (3) the mediator variable is independently and significantly correlated with the dependent variable, and (4) the independent variable drops in strength and/or significance when the independent and mediator variables are both entered into the regression equation.

Results

Descriptive statistics and measurement models

Table 1 indicates means and standard deviations for each scale and interscale correlations for all study variables. As shown in Table 1, the interscale correlations were all significant ($p < 0.01$) in the expected direction of association. Multicollinearity was not a problem in this study, since correlations were below the 0.70 threshold (Fidell et al. 1996). In addition, all variance inflation factors in the regressions were <2, suggesting that all measures were appropriate for inclusion in the analyses (Fidell et al. 1996).

In order to detect and control for the possible influence of common method bias statistically, we performed confirmatory factor analyses (CFA) on the data set. Specifically, we calculated six fit indices – χ^2 / df, Tucker-Lewis, incremental fit, comparative fit, root mean square error of approximation, and standardised root mean square residual – to determine how the model fitted our data. A good fit is determined by values <2.5, >0.9, >0.9, >0.9, <0.08, and <0.08, respectively (Arbuckle 2006; Hu and Bentler 1998). A series of nested model comparisons – comparing the full measurement model comprising all latent variables with a range of alternative models – suggested that all variables in this study were distinct and thus appropriate for inclusion in the analyses (Table 2). The fit of the intended model with four distinct variables was significantly better ($p < 0.001$) than all other models.

Furthermore, the measurement model exhibited good psychometric properties (Table 2), with all items loading onto their latent factors (Anderson and Gerbing 1988). The path estimates were all significant ($p < 0.001$), with the lowest standardised value being 0.45. We applied Harman's single-factor test as an additional check for common method variance; when all variables were allowed to load onto one general factor, that did not account for most of the variance in the data.

Test of hypotheses

As shown in Table 3, results in Step 1 of the first column (i.e. when the dependent variable was innovativeness) indicated that the control variables of organisational size and age were not statistically significantly related to innovativeness ($\beta = .04$, $p > .05$ and $\beta = .06$, $p > .05$, respectively). Results in Step 2 of the first column (i.e. when the dependent variable was innovativeness) indicated that corporate foresight was significantly related to innovativeness ($\beta = .24$, $p < .01$), thereby supporting H1 and satisfying the first condition for mediation. Step 3 of the first column (i.e. when the dependent variable was innovativeness) further revealed that organisational learning was significantly related to innovativeness ($\beta = .31$, $p < .01$), and Step 2 of the second column (i.e. when the dependent variable was organisational learning) showed that corporate foresight was significantly related to organisational learning ($\beta = .41$, $p < .01$). Therefore, H2 was supported, and the next two requirements of mediation were met. When entering corporate foresight and organisational learning into the model simultaneously, corporate foresight was no longer significant ($\beta = .08$, $p > .05$). This result indicates that organisational learning mediates the relationship between corporate foresight and innovativeness, thus supporting H3.

To assess moderated mediation, Hypothesis 4 was addressed in a stepwise fashion. Specifically, we first needed to demonstrate that corporate foresight was associated with innovativeness. Step 2 of the first column (i.e. when the dependent variable was innovativeness) in Table 3 presents this

Table 1. Descriptive statistics, correlations, and scale reliabilities of study variables.

		Internal consistency	Mean	SD	1	2	3
1	Corporate foresight	0.82	4.32	1.21			
2	Organisational learning	0.88	5.65	0.67	0.37**		
3	Integrative capabilities	0.92	5.42	1.17	0.42**	0.41**	
4	Innovation	0.78	4.34	1.56	0.39**	0.31**	0.30**

Notes: $n = 212$, $^{**}p < 0.01$.

Table 2. Measurement model comparisons.

Models	χ^2 (df)	TLI	IFI	CFI	RMSEA	SRMR	χ^2_{diff}
Full measurement model, four-factor	764 (338)	0.925	0.943	0.952	0.058	0.052	
Model A, three factors[a]	897 (343)	0.878	0.899	0.902	0.082	0.076	133***
Model B, three factors[b]	1501 (343)	0.774	0.811	0.788	0.124	0.122	737***
Model C, three factors[c]	1343 (343)	0.812	0.828	0.826	0.094	0.112	579***
Model D, two factors[d]	1885 (347)	0.754	0.776	0.754	0.116	0.102	1121***
Model E, one factor[e]	2746 (353)	0.532	0.584	0.588	0.154	0.132	1982***

Notes: $n = 212$, ***$p < 0.001$; χ^2, chi-square discrepancy; df, degrees of freedom; TLI, Tucker-Lewis Index; IFI, incremental fit index; CFI, comparative fit index; RMSEA, root mean square error of approximation; SRMR, standardised root mean square residual; χ2, difference in chi-square.

[a]Corporate foresight and organisational learning combined into a single factor; compared to full measurement four-factor model.

[b]Corporate foresight and integrative capabilities combined into a single factor; compared to full measurement four-factor model.

[c]Organizational learning and integrative capabilities combined into a single factor; compared to full measurement four-factor model.

[d]Corporate foresight, organisational learning, and integrative capabilities combined into a single factor; compared to full measurement four-factor model.

[e]Harman's single factor model; all variables combined into a single factor; compared to full measurement four-factor model.

Table 3. Hierarchical regression results for testing mediation.

	DV = Innovation				DV = Organisational learning	
Variable and statistic	Step 1	Step 2	Step 3	Step 4	Step 1	Step 2
The size of organisation	0.04	0.06	0.04	0.03	0.06	0.06
The age of organisation	0.09	0.11	0.12	0.11	0.09	0.05
Corporate foresight		0.24**		0.08		0.41**
Organisational learning			.31**	0.32**		
F	0.5	5.34**	6.52**	5.89**	0.4	9.61**
R^2	0.03	0.15**	0.19	0.20**	0.06	0.28**
(Adjusted R^2)	(0)	(0.12)	(0.16)	(0.17)	(0.04)	(0.25)

Notes: $n = 212$, **$p < 0.01$; Control variables include the size of organisation and the age of organisation. Significance testing of R^2 is compared to the control model.

significant result ($\beta = .24$, $p < .01$). Second, we needed to see if there was a significant interaction between corporate foresight and the proposed moderator of integrative capabilities on innovativeness. Results revealed that the moderator of integrative capabilities interacted with corporate foresight to be positively related to organisational learning ($\beta = .15$, $p < .01$; Table 4), supporting H4.

Discussion

In response to calls for analysis of the spurious relationship between corporate foresight and innovativeness (e.g. Rohrbeck and Gemünden 2011), we explored processes of mediation and moderation linking corporate foresight to innovativeness. Specifically, we proposed that organisational learning mediates the relationship between corporate foresight and innovativeness and that the strength of the association between corporate foresight and organisational learning is moderated by an

Table 4. Regression results for testing moderation of integrative capabilities on the relationship between corporate foresight and organisational learning.

		Organisational learning		
Step	Variable	B	R^2 (Adjusted R^2)	F
1	Control variables		0.06 (0.04)	0.40
2	Corporate foresight	0.41**	0.28 (0.25)**	9.61**
	Integrative capabilities	0.05*		
3	Corporate foresight X Integrative capabilities	0.15**	0.29 (0.26)**	9.74**

Notes: $n = 212$, *$p < 0.05$, **$p < 0.01$; Control variables include the size of organisation and the age of organisation. Significance testing of R^2 for Step 3 is compared to the Step 2.

organisation's integrative capabilities. Our theoretical model was supported by statistical analyses based on data from 212 manufacturing organisations in South Korea. The results have several theoretical and practical implications.

Theoretical implications

This study sheds light on the nature of the relationship between corporate foresight and innovativeness. Building on RBV and the literature of corporate foresight, we theorised and examined the positive relationship between corporate foresight and innovativeness because corporate foresight is an organisational resource that enables the company to address external changes in a timely manner to ensure the long-term survival and success of the company (Rohrbeck 2010a). Our study results indicate that with more corporate foresight exercises, there are high levels of innovativeness. The finding adds to a growing collection of corporate foresight studies that examine the effect of corporate foresight on an organisation's innovation capacity (e.g. Von der Gracht, Vennemann, and Darkow 2010).

Next, our research illuminates organisational learning as a mechanism that accounts for the relationship between corporate foresight and innovativeness. Organisations that employ more corporate foresight exercises are more likely to be engaged in learning and, therefore, have a higher propensity to enact innovation. Previous research focused mainly on the direct relationship between corporate foresight and innovation (e.g. Anderson et al. 2004). By introducing a new mediator in that relationship, we argue that organisational learning is a specific channel through which corporate foresight influences innovativeness. Corporate foresight is viewed through RBV as an intangible resource that influence organisational learning.

Finally, we showed that a contextual variable impacts the process through which corporate foresight influences organisational learning and therefore innovativeness. The extent to which high levels of corporate foresight lead to high levels of organisational learning and thus innovativeness is dependent on the integrative capabilities of an organisation. Indeed, through an organisation's capacity to combine and consolidate knowledge and skills, corporate foresight might lead the organisation to solve problems and experiment (Yoon and Kayes 2016). This study's findings have important implications for the literatures of corporate foresight and organisational learning, which have not fully explained the conditions under which organisational learning and corporate foresight lead to innovativeness.

Practical implications

Our results indicate that organisational leaders should be aware that, in addition to organisational learning, a contextual variable of integrative capabilities is critical to establishing a positive environment in which high levels of corporate foresight lead to high levels of organisational learning and thus innovativeness. Although there is a relationship between high levels of corporate foresight and high levels of organisational learning, high levels of integrative capabilities are needed to maximise the benefit of corporate foresight for innovativeness.

Many organisations find it difficult to adapt quickly enough in times of discontinuous change and can find themselves in life threatening situations (Stubbart and Knight 2006). An integrative capability to combine and process knowledge from multiple, distributed sources within their boundaries and beyond helps organisations address the challenges of developing a new product or a new process innovation.

Arguably, the objective of technology and innovation management is to integrate knowledge from external sources, blending the different technical competencies required for the task. The crucial question for organisational leaders is how to manage this process of integration to achieve those objectives. Our results show that the extent to which corporate foresight is translated into high levels of organisational learning and innovativeness varies as a function of the capability of the organisation to integrate knowledge and skills. Creating a highly learning workforce has become a significant focus for many organisations (MacLeod and Clarke 2009), and our study shows that organisational leaders and members have to work together to create a virtuous cycle to ensure that organisations with high levels of corporate foresight demonstrate organisational learning and are likely to integrate knowledge to enhance innovation.

Limitations and future studies

Since this was a nonexperimental study, it was not designed to determine causality. Nevertheless, there is a reasonable level of confidence in the study's results, particularly since the study was rooted in the theoretical bases of corporate foresight, RBV, and dynamic capabilities theory. Future research may want to employ an experimental, longitudinal, or panel design to analyze how different levels of corporate foresight, organisational learning, and integrative capabilities impact different levels of innovation in organisations.

Future research may also want to consider the relationship between innovativeness and innovation. Although the relationship was beyond the scope of this paper, future study can extend the current study by investigating whether innovativeness, enhanced by corporate foresight, leads to innovation measured by performance outcome.

Conclusion

Our study explored the complex relationship between corporate foresight and innovation. Based on the statistical analyses, we found that organisational learning mediated the relationship between corporate foresight and innovativeness and that the organisation's integrative capabilities moderated the relationship between corporate foresight and organisational learning; thus, the indirect effect of corporate foresight on innovation through organisational learning was conditional on the organisation's level of integrative capabilities. The findings are consistent with the central propositions of RBV and dynamic capabilities theory and provide empirical evidence that corporate foresight, organisational learning, and integrative capabilities are important for enhancing innovation.

Disclosure statement

No potential conflict of interest was reported by the authors.

Funding

This work was supported by a Korea University Grant.

References

Aiken, L. S., S. G. West, and R. R. Reno. 1991. *Multiple Regression: Testing and Interpreting Interactions*. Thousand Oaks, CA: Sage.

Ambrosini, V., and C. Bowman. 2009. "What Are Dynamic Capabilities and Are They a Useful Construct in Strategic Management?" *International Journal of Management Reviews* 11: 29–49.

Anderson, J. C., and D. W. Gerbing. 1988. "Structural Equation Modeling in Practice: A Review and Recommended Two-step Approach." *Psychological Bulletin* 103 (3): 411–423.

Anderson, P. D., B. H. Jorgensen, L. Lading, and B. Rasmussen. 2004. "Sensor Foresight—Technology and Market." *Technovation* 24: 311–320.

Arbuckle, J. L. 2006. *Amos (Version 7.0) [Computer Program]*. Chicago: SPSS.

Arnold, H. M. 2003. *Technology Shocks: Origins, Management Responses and Firm Performance*. Heidelberg: Physica Verlag Springer.

Barney, J. B. 1986. "Strategic Factor Markets: Expectations, Luck, and Business Strategy." *Management Science* 32: 1231–1241.

Barney, J. B. 1991. "Firm Resources and Sustained Competitive Advantage." *Journal of Management* 17: 99–120.

Baron, R. M., and D. A. Kenny. 1986. "The Moderator–Mediator Variable Distinction in Social Psychological Research: Conceptual, Strategic, and Statistical Considerations." *Journal of Personality and Social Psychology* 51: 1173–1182.

Battistella, C. 2014. "The Organisation of Corporate Foresight: A Multiple Case Study in the Telecommunication Industry." *Technological Forecasting and Social Change* 87: 60–79.

Burke, J. W. 1989. *Competency Based Education and Training*. New York: Psychology Press.

Christie, A. A. 1990. "Aggregation of Test Statistics: An Evaluation of the Evidence on Contracting and Size Hypotheses." *Journal of Accounting and Economics* 12 (1): 15–36.

Cohen, W. M., and D. A. Levinthal. 1990. "Absorptive Capacity: A New Perspective on Learning and Innovation." *Administrative Science Quarterly* 35: 128–152.

Collis, D., and C. Montgomery. 1995. "Competing on Resources: Strategy in the 1990s." *Harvard Business Review* 73: 118–128.

Damanpour, F. 1991. "Organizational Innovation: A Meta-analysis of Effects of Determinants and Moderators." *Academy of Management Journal* 34: 555–590.

Edmondson, A. 1999. "Psychological Safety and Learning Behavior in Work Teams." *Administrative Science Quarterly* 44: 350–383.

Eisenhardt, K. M., and J. A. Martin. 2000. "Dynamic Capabilities: What Are They?" *Strategic Management Journal* 21: 1105–1121.

Fidell, S., L. Silvati, R. Howe, K. S. Pearsons, B. Tabachnick, R. C. Knopf, and T. Buchanan. 1996. "Effects of Aircraft Overflights on Wilderness Recreationists." *The Journal of the Acoustical Society of America* 100: 2909–2918.

Firebaugh, G. 1978. "A Rule for Inferring Individual-level Relationships From aggregate Data." *American Sociological Review* 43: 557–572.

Georghiou, L., and J. C. Harper. 2011. "From Priority-setting to Articulation of Demand: Foresight for Research and Innovation Policy and Strategy." *Futures* 43: 243–251.

Gershman, M., S. Bredikhin, and K. Vishnevskiy. 2016. "The Role of Corporate Foresight and Technology Roadmapping in Companies" Innovation Development: The Case of Russian State-owned Enterprises." *Technological Forecasting and Social Change* 110: 187–195.

Griffith, T. L., and J. E. Sawyer. 2010. "Multilevel Knowledge and Team Performance." *Journal of Organizational Behavior* 31: 1003–1031.

Helfat, D. E., and M. A. Peteraf. 2003. "The Dynamic Resource-based View: Capability Lifecycles." *Strategic Management Journal* 24: 997–1010.

Helfat, C. E., and R. S. Raubitschek. 2000. "Product Sequencing: Co-evolution of Knowledge, Capabilities and Products." *Strategic Management Journal* 21: 961–979.

Hitt, M. A., R. D. Nixon, P. G. Clifford, and K. P. Coyne. 1999. "The Development and Use of Strategic Resources." In *Dynamic Strategic Resources: Development, Diffusion and Integration*, edited by M. A. Hitt, P. G. Clifford, R. D. Nixon, and K. P. Coyne, 1–15. New York: Wiley.

Hmieleski, K. M., and R. A. Baron. 2009. "Entrepreneurs" Optimism and New Venture Performance: A Social Cognitive Perspective." *Academy of Management Journal* 52: 473–488.

Hu, L. T., and P. M. Bentler. 1998. "Fit Indices in Covariance Structure Modeling: Sensitivity to Underparameterized Model Misspecification." *Psychological Methods* 3 (4): 424–453.

Huber, G. P. 1991. "Organizational Learning: The Contributing Processes and the Literatures." *Organization Science* 2: 88–115.

Hurley, R. F., and G. T. M. Hult. 1998. "Innovation, Market Orientation, and Organizational Learning: An Integration and Empirical Examination." *Journal of Marketing* 62 (3): 42–54.

Khanna, T., R. Gulati, and N. Nohria. 1998. "The Dynamics of Learning Alliances: Competition, Cooperation, and Relative Scope." *Strategic Management Journal* 19 (3): 193–210.

Khripunova, A., K. Vishnevskiy, O. Karasev, and D. Meissner. 2014. "Corporate Foresight for Corporate Functions: Impacts from Purchasing Functions." *Strategic Change* 23: 147–160.

Kogut, B., and U. Zander. 1992. "Knowledge of the Firm, Combinative Capabilities, and the Replication of Technology." *Organization Science* 3: 383–397.

Lichtenthaler, U. 2009. "Absorptive Capacity, Environmental Turbulence, and the Complementarity of Organizational Learning Processes." *Academy of Management Journal* 52: 822–846.

Lorenzoni, G., and A. Lipparini. 1999. "The Leveraging of Inter-firm Relationships as a Distinctive Organizational Capability: A Longitudinal Study." *Strategic Management Journal* 20: 317–338.

MacLeod, D., and N. Clarke. 2009. *Engaging for Success: Enhancing Performance Through Employee Engagement: A Report to Government*. London: Department for Business, Innovation and Skills.

May, R. C., W. H. Stewart, and R. Sweo. 2000. "Environmental Scanning Behavior in a Transitional Economy: Evidence from Russia." *Academy of Management Journal* 43 (3): 403–427.

Mothe, C., and B. V. Quelin. 1998. "How Firms Can Benefit from Collaborating Within R&D Consortia." In *Managing Strategically in an Interconnected World*, edited by M. Hitt, J. Ricart, and R. Nixon, 321–348. New York: Wiley.

Nadler, D., M. Tushman, and M. B. Nadler. 1997. *Competing by Design: The Power of Organizational Architecture*. New York: Oxford University Press.

Podsakoff, P. M., S. B. MacKenzie, J. Y. Lee, and N. P. Podsakoff. 2003. "Common Method Biases in Behavioral Research: A Critical Review of the Literature and Recommended Remedies." *Journal of Applied Psychology* 88 (5): 879–903.

Preacher, K. J., D. D. Rucker, and A. F. Hayes. 2007. "Addressing Moderated Mediation Hypotheses: Theory, Methods, and Prescriptions." *Multivariate Behavioral Research* 42: 185–227.

Ramírez, R., L. Roodhart, and W. Manders. 2011. "How Shell's Domains Link Innovation and Strategy." *Long Range Planning* 44 (4): 250–270.

Rohrbeck, R. 2010a. *Corporate Foresight: Towards a Maturity Model for the Future Orientation of a Firm*. New York: Springer.

Rohrbeck, R. 2010b. "Harnessing a Network of Experts for Competitive Advantage: Technology Scouting in the ICT Industry." *R&D Management* 40 (2): 169–180.

Rohrbeck, R. 2012. "Exploring Value Creation from Corporate-foresight Activities." *Futures* 44 (5): 440–452.

Rohrbeck, R., and H. G. Gemünden. 2011. "Corporate Foresight: Its Three Roles in Enhancing the Innovation Capacity of a Firm." *Technological Forecasting and Social Change* 78 (2): 231–243.

Rothaermel, F. T. 2001. "Incumbent's Advantage Through Exploiting Complementary Assets Via Interfirm Cooperation." *Strategic Management Journal* 22 (6-7): 687–699.

Ruff, F. 2015. "The Advanced Role of Corporate Foresight in Innovation and Strategic Management—Reflections on Practical Experiences from the Automotive Industry." *Technological Forecasting and Social Change* 101: 37–48.

Russo, M. V., and P. A. Fouts. 1997. "A Resource-based Perspective on Corporate Environmental Performance and Profitability." *Academy of Management Journal* 40 (3): 534–559.

Sarpong, D., and M. Maclean. 2012. "Mobilising Differential Visions for New Product Innovation." *Technovation* 32 (12): 694–702.

Sarpong, D., M. Maclean, and E. Alexander. 2013. "Organizing Strategic Foresight: A Contextual Practice of 'Way Finding.'." *Futures* 53: 33–41.

Shrivastava, P., and J. H. Grant. 1985. "Empirically Derived Models of Strategic Decision-Making Processes." *Strategic Management Journal* 6 (2): 97–113.

Spanos, Y. E., N. S. Vonortas, and I. Voudouris. 2015. "Antecedents of Innovation Impacts in Publicly Funded Collaborative R&D Projects." *Technovation* 36-37: 53–64.

Stubbart, C. I., and M. B. Knight. 2006. "The Case of the Disappearing Firms: Empirical Evidence and Implications." *Journal of Organizational Behavior* 27 (1): 79–100.

Teece, D. J., G. Pisano, and A. Shuen. 1997. "Dynamic Capabilities and Strategic Management." *Strategic Management Journal* 18: 509–533.

Thorpe, R., R. Holt, A. Macpherson, and L. Pittaway. 2005. "Using Knowledge Within Small and Medium-Sized Firms: A Systematic Review of the Evidence." *International Journal of Management Reviews* 7 (4): 257–281.

Tushman, M. L., B. Virany, and E. Romanelli. 1985. "Executive Succession, Strategic Reorientations, and Organization Evolution: The Minicomputer Industry as a Case in Point." *Technology in Society* 7 (2-3): 297–313.

Van de Ven, A. H., and H. L. Angle. 1989. *Suggestions for Managing the Innovation Journey (No. 9)*. Minneapolis, MN: Strategic Management Research Center, University of Minnesota.

Vecchiato, R., and C. Roveda. 2010. "Strategic Foresight in Corporate Organizations: Handling the Effect and Response Uncertainty of Technology and Social Drivers." *Technological Forecasting and Social Change* 77: 1527–1539.

Vishnevskiy, K., O. Karasev, and D. Meissner. 2015. "Integrated Roadmaps and Corporate Foresight as Tools of Innovation Management: The Case of Russian Companies." *Technological Forecasting and Social Change* 90 (B): 433–443.

Von der Gracht, H. A., C. R. Vennemann, and I. Darkow. 2010. "Corporate Foresight and Innovation Management: A Portfolio-approach in Evaluating Organisational Development." *Futures* 42 (4): 380–393.

Waldman, D. A., and B. M. Bass. 1991. "Transformational Leadership at Different Phases of the Innovation Process." *The Journal of High Technology Management Research* 2 (2): 169–180.

Yam, R. C. M., J. C. Guan, K. F. Pun, and E. P. Y. Tang. 2004. "An Audit of Technological Innovation Capabilities in Chinese Firms: Some Empirical Findings in Beijing, China." *Research Policy* 33 (8): 1123–1140.

Yoon, J., and D. C. Kayes. 2016. "Employees' Self-efficacy and Perception of Individual Learning in Teams: The Cross-level Moderating Role of Team-learning Behavior." *Journal of Organizational Behavior* 37 (7): 1044–1060.

Corporate foresight in open laboratories – a translational approach

Albrecht Fritzsche

ABSTRACT

Many companies have lately started to use open laboratories for innovation. This paper looks into the added value of these practices for corporate foresight. To differentiate corporate foresight in open laboratories from innovation, the paper uses the concept of translation from cultural theory, which addresses the treatment of semantic differences in the ways how different groups of people express meaning. Based on a multiple case study, the paper collects evidence about translation practices in open laboratories and investigates how they relate to corporate foresight. The results show that translation in open laboratories enables companies to gain insight into their own situation and engage in an ongoing re-negotiation of their relation with other stakeholders in commercial interactions. This is a distinctive aspect of corporate foresight in open laboratories which has so far been widely neglected in research. It creates various opportunities for further development of foresight practices.

Introduction

In the dynamic and volatile economic environment of the early twenty-first century, corporate foresight and innovation management have developed a very close relationship. According to von der Gracht, Vennemann, and Darkow (2010), one can even speak about a symbiosis between the two functions: innovation management increasingly assumes a strategic role in setting the direction for further development by designing business models (Massa and Tucci 2013) and establishing ecosystems for shared value creation (Gawer and Cusumano 2014); at the same time, corporate foresight engages in common activities with decision makers using foresight data (Cuhls 2003; Rohrbeck and Gemünden 2011) which provide new opportunities for immediate learning and direct validation of hypotheses about future trends in practice (Daheim and Uerz 2008; Rohrbeck, Battistella, and Huizingh 2015). The overall goal of corporate foresight is the early detection of economic change to ensure the strategic fit of the company to its environment (Battistella and De Toni 2011; Ruff 2015). It is therefore particularly interesting for corporate foresight to engage in activities which involve external participants to gather information about the company's environment right from the source. Open laboratories provide a perfect setting to do so.

Many companies have recently become involved in open laboratories. The Ford Motor Company, for example, works with a TechShop which gives access to advanced fabrication tools for everyone who wants to build objects and devices (Richardson, Elliott, and Haylock 2013). Ford encourages employees to use the TechShop, start tinkering and exchange with other enthusiasts from inside and outside the company in idea generation and solution development. BMW pursues similar goals with a Maker Space (Troxler 2016). GE Appliances has started its open microfactory 'FirstBuild'; Companies like Airbus, Chevron and Northrop Grumman have set up Fab Labs for their employees (Troxler 2015). Ericsson and others have turned towards Living Labs (Bergvall-Kåreborn and Ståhlbröst 2009), pursuing a less technology-centric and more human-centric approach to collaboration (Levén and Holmström 2008). Lego and German Telekom have established interactive spaces in store settings to engage with customers in the exploration of new products and services (Roth et al. 2015).

Scientific literature emphasises the added value of collaborative design and information exchange in open laboratories for innovation (Almirall, Lee, and Wareham 2012; Hatch 2014; Troxler 2015). Open laboratories, however, also have other advantages for companies: they include them in collective urban governance (Voytenko et al. 2016), connect them to larger social or political movements (Toupin 2014), support training and education (Halverson and Sheridan 2014) and mediate technology transfer (De Arias et al. 2014). Open laboratories are therefore interesting for other corporate functions than innovation management as well. As Hyysalo et al. (2014) have shown, open laboratories can be used as settings for trend identification; furthermore, they are known to support grass-root strategizing activities, particularly in the context of urban sustainability (e.g. Boeri et al. 2017). This raises the question about the contribution of open laboratories to corporate foresight which is addressed on the following pages.

Instead of looking at the application of specific foresight techniques in open laboratories, this paper picks up on the notion of a symbiotic relationship between corporate foresight and innovation management (von der Gracht, Vennemann, and Darkow 2010). It explores how the engagement with external contributors during innovation activities in open laboratories generates insights about the company which are relevant for corporate foresight. As a theoretical basis, the paper uses recent work on cultural translation, which posits that the experience of contrast and contradiction is essential to the formation of any social relationship. Evidence from a multiple case study indicates that this process also applies to the creation of a strategic fit between companies and their economic environment, which opens a new dimension for the discussion of corporate foresight.

Theoretical background

Corporate foresight as an open, interventional activity

The origins of the current discussion on corporate foresight date back to the middle of the twentieth century (Jefferson 2012; Rohrbeck, Battistella, and Huizingh 2015). Nevertheless, many aspects of the practice of corporate foresight, particularly in relation of other strategic activities, are still little understood (von der Gracht, Vennemann, and Darkow 2010; Rohrbeck and Gemünden 2011; Rhisiart, Miller, and Brooks 2015). Daheim and Uerz (2008) distinguish four different stages in the development of strategic foresight exercises: from expert-based foresight over the usage of models and trend calculations towards the currently expanding practice of context-based open foresight activities, which 'pays tribute to the increased socio-cultural and socio-technical dynamic resulting from the emergence of the networked society' and focusses on open communication between all relevant stakeholders, which is 'set to diffuse into decision-making and blend into it instead of just preparing it' (Daheim and Uerz 2008, 13). This form of corporate foresight corresponds with a stronger interaction between companies and other stakeholders in open innovation activities (von der Gracht, Vennemann, and Darkow 2010).

Rohrbeck, Battistella, and Huizingh (2015) take another approach at the distinction of different phases in the development of corporate foresight, starting with its early attachment to systems thinking, followed by scenario planning in the 1960s and 1970s, the elaboration of methods and processes in the 1980s and 1990s, up to the understanding of corporate foresight as an general organisational ability today. Rohrbeck and Gemünden (2011) accordingly propose to assess corporate foresight exercises regarding their general contribution to the dynamic capabilities of a company in changing business environments.

The development of corporate foresight over the decades goes along with a wider change in business and management studies regarding the empirical methods applied in scientific research. Since the 1970s, qualitative methods have steadily gained importance which turn research into 'a situated activity that locates the observer in the world' (Denzin and Lincoln 2005, 3). Analytic approaches which try to uphold an external perspective upon the observed phenomena are complemented with hands-on activities which are more closely related to bricolage than systematic problem solving (Boxenbaum and Rouleau 2011; Klag and Langley 2013). Recent practices in corporate foresight and strategy can be described in a similar way (Cunha, Cunha, and Clegg 2008). Corporate foresight thus becomes deeply entangled with a variety of different function is a company (Rhisiart, Miller, and Brooks 2015): innovation management is one of them, next to other strategic decision-making and business development processes. Recent literature accordingly discusses the relation of corporate foresight to innovation in a general manner without assuming direct effects on specific outcomes (Anderson et al. 2004; von der Gracht, Vennemann, and Darkow 2010; Ramirez, Roodhart, and Manders 2011). Corporate foresight is rather expected to create the necessary 'new mindset' (Ratcliffe 2005), based on a 'holistic approach' (Durst et al. 2015), and an assessment of the company's 'strategic fit with the future' (Battistella and De Toni 2011).

Open laboratories

The conditions for corporate foresight practices in open laboratories are radically different from conventional settings. Company experts who engage with external participants in common activities do not only gain insight and directions for corporate development. They put them into action at the same time as they build the foundations for future business operations together with customers and other stakeholders on the spot. It is this simultaneity that needs to be better understood in order to describe the special value of open laboratories for corporate foresight.

Open laboratories are closely related to the concepts of open innovation, which purposively makes use of knowledge flows across organisational boundaries (Chesbrough 2003), and user innovation, which turns the focus to the inventiveness of people in coming up with solutions to their own problems in daily business (von Hippel 2009). Innovation in open laboratories can be regarded as the outcome of the interaction in holistic business ecosystems (Adner and Kapoor 2010). Companies have lately spent particular effort on the integration of customers in their innovation activities, using virtual settings on the internet (Rohrbeck, Steinhoff, and Perder 2010). Such settings include discussion groups, marketplaces for ideas and solutions, virtual design studios and other types of tools, which can be combined in different ways (Möslein and Fritzsche 2017). Open laboratories provide similar settings in an offline environment, where people can meet in person. This gives them further opportunities to interact without involvement of information technology.

Open laboratories can be designed and equipped quite differently. FabLabs and TechShops focus on the construction of technical artefacts. They contain large sets of tools and machines like 3d printers, laser cutters, etc. Which enable easy prototyping of solutions (Gershenfeld 2008; Hatch 2014). Living Labs do not necessarily use advanced technical equipment. They focus on open exchange and networking to engage people in common activities (Leminen, Westerlund, and Nyström 2012). Labs like these revolve around a wider notion of collective effort and communities as drivers for innovation (Benkler 2011; Fischer 2013).

Different types of open laboratories can also be distinguished in other dimensions. With O'Hern and Rindfleisch (2008), one can differentiate between various types of interaction: interaction with and without predetermined types of contribution, and interaction lead by consumers or by producers. Roth et al. (2015) distinguish laboratories run by companies from those run by intermediaries, and permanent installations from event-based setups. In any case, open laboratories facilitate a transgression of industrial boundaries in the practice of innovation, which forces professionals to abandon accustomed procedural patterns, role models and hierarchical structures (Fritzsche 2017). Corporate experts are permanently required to set the company's interests in relation to the interests of other stakeholders who perceive common business operations from a different point of view. Innovation activities in open laboratories are therefore always accompanied by a search for a common understanding of business and a shared meaning of value, which allows companies to learn more about its own position on the market and future directions for development. The concept of cultural translation makes it possible to describe this process in more detail.

Cultural translation and translation spaces

Cultural translation studies have long departed from the simplistic understanding of translation as a substitution exercise, replacing expressions in one language by others in a different one. Since the use of each language is embedded in cultural practice, finding direct equivalents for words of two languages it is generally impossible (Bassnett 2013). According to Benjamin (1968), the work of the translator is therefore a creative achievement in its own way, because it involves choices how to connect cultures: what has meaning in one culture is related to something else which has meaning in another one. Activities among the members of one culture can thus resonate in other activities among members of a second culture (cf. Rosa 2010). Cultural translation studies explore how such resonance is achieved. This includes two different aspects.

On the background of actor-network theory, translation can be studied as a process in which 'the identity of actors, the possibility of interaction and the margins of manoeuvre are negotiated and delimited' (Callon 1986, 203). The research interest is focussed on the conclusions actors draw from translation about their mutual relationship. This includes relationships between functional domains, which are considered to be bound together by 'delegation' (Latour 1990): they 'co-constitute' themselves in complementary roles.

Post-colonial theories rather focus on the provisional nature of such relationships: the notions which actors achieve about themselves and others through translation are never fully accurate and exhaustive; they constantly need to be questioned, confirmed or adapted. One could say that translation is the opposite of a rational problem solving process with a clear starting and end point, because there is no ultimate resolution of the conflicts between the parties involved. Translation therefore seems applicable a concept to explain ongoing activities like corporate foresight, which continuously questions and reviews the position of a company and its future strategy.

With Foucault (1971), spaces in which translation proceeds can be described as heterotopias: real-world settings where contrasts and contradictions between different societal or personal interests and perspectives are acted out. Heterotopias allow the experience of different layers of meaning and appearance in certain phenomena as they really are, without looking for consistency and integration. While the concept of heterotopias has not remained unchallenged (Genocchio 1995; Saldanha 2008), it has been highly influential in the spatial turn in social theory (Warf and Arias 2008), which can be considered as a reaction to the ubiquity of systemic models at the end of the twentieth century. While systemic models assume that people are, in one way or another, functional parts of a superstructure, spatial models don't. They focus on differences between social groups or individuals which can ultimately not be resolved, but involve a 'third space' where transformations and transgressions from one to another take place without annihilating individual identities (Bhabha 2004). This third space is considered essential for societal development, as it allows exploration and formation of novelty in-between fixed cultural identities (Rutherford 1990).

Architectural studies have already discussed the design of such spaces quite extensively in urban planning (Dehaene and De Cauter 2008), regarding parks, assembly halls, marketplaces, theatres, but also for virtual settings of interaction (Liff 2003). An application to open laboratories seems justified, too, since they also attract people with common interests but different ways of expressing themselves (Gershenfeld 2008; Westerlund and Leminen 2011; McPhee, Westerlund, and Leminen 2012; Hatch 2014). In contrast to knowledge-based approaches, heterotopic or spatial approaches to open laboratories are less interested in resource integration. They turn the focus to the experience of contradiction and conflict as a background of economic interaction and the ongoing negotiation of relationships between the stakeholders. Open laboratories are, in this sense, no 'production facilities' for innovation and knowledge. Their impact cannot be reduced to a series of appropriable output items for a company. Companies can benefit from the engagement in open laboratories on a wider scope, which relates directly to the understanding of corporate foresight as an open, dynamic activity of bricolage. Open laboratories allow companies to reconfirm and actualise their own position within their current environment and to set directions for further activity.

Research design

Methodological approach

The empirical study presented on the following pages takes a closer look at the practice of cultural translation in open laboratories to understand their actual contribution to corporate foresight. Since the concept of translation has not been used in this context before, an empirical method is necessary which complements incremental theory building with practical evidence. According to (Eisenhardt 1989, 548–549) case studies serve this purpose very well. Scholars in the field of corporate foresight often use case studies in order to gain a sound qualitative understanding of the phenomena which are investigated (von der Gracht, Vennemann, and Darkow 2010; Rohrbeck and Gemünden 2011). Case study research also plays an important role in the context of open innovation, 'as it increases our understanding of how things work and it enables us to identify important concepts and phenomena, e.g. effective open innovation practices and crucial context characteristics' (Huizingh 2011, 7).

Yin (2009) differentiates singe, multiple, holistic and embedded case study designs. Single case study designs are considered to be particularly suitable for critical or extreme cases and the testing of a well-established theory, while multiple designs increase the methodological rigour of the investigation by 'strengthening the precision, the validity and stability of the findings,' (Miles and Huberman 1994, 30). Holistic designs consider a specific context for each case. Embedded designs study entities within one and the same context to gain insight into different facets of the overall phenomenon (Yin 2009). The subject matter at hand suggests a multiple, embedded case study design which looks at a set of different activities performed in an open laboratory. This makes it possible to reduce biases caused by specific topics addressed in these activities. At the same time, the common context of the open laboratory ensures comparability of the findings and reduces disturbances by external influences.

The aim of the study is to identify translation practices in open laboratories which go beyond the scope of the problem solving activities in innovation management and to find out how they affect corporate foresight. As mentioned before, such translations proceed through the experience of conflicts and differences between the parties involved and the insights they gain from this experience about themselves and their relationship to other parties. Conflicts and differences can surface during the interaction in open laboratories on two levels: one concerns the content of the contributions which are made; the other concerns the way how the contributors interact, how they take the statements of others into consideration, build on them or discard them when they add something from their own point of view. This makes it necessary to take a multimodal qualitative approach in the analysis of the case units to assess content and interactions of the contributors at the same time. The case study presented here therefore combines techniques of content analysis (Mayring 2000) and conversation analysis (Wooffitt 2005) on different data collected from direct observations, interviews with the personnel in the laboratories and original material produced during the interaction of the participants in the form of text or physical artefacts.

Case unit selection

The open laboratory in which the data were collected is situated in the centre of a major European city in direct vicinity of major department stores, electronic retailers and many other shops and restaurants. It is run by an independent research institute with a regular staff of four persons on location, supported by numerous students who welcome visitors during normal shopping hours. A coffee shop is part of the design. The laboratory includes a workbench with technical infrastructure and a meeting room with extensive material for interactive sessions, used by companies for internal activities and by various other institutions for public events. While there is also a small shopping area included, most of the space is dedicated to projects of companies, public institutions or research centres. These projects use the laboratory over a period of three months to exchange with visitors who enter the premises whenever they want and with whatever intentions they have in mind.

The laboratory was chosen because it offers a broad range of opportunities for companies to interact with visitors. The technical equipment in the laboratory plays generally a minor role, although companies are allowed to setup specific devices which can be used by the visitors. Furthermore, there are no restrictions on the type of engagement with the customers. Companies use the laboratory for a large number of different topics throughout the whole innovation process from idea generation to prototype testing. In the early stages, drawings or mock-ups inspire exchange with customers about idea generation. In later stages, functional prototypes and pilot application scenarios are used to discuss the feasibility and value of the solutions. Independently from the subject of their interaction with the companies, visitors are free to test, tinker around and explore alternatives. Depending on the setup chosen by the company, they can leave input in the form of written comments, designed artefacts or conversations with the shop personnel and company representatives.

Table 1. Subjects of interaction in the open laboratory.

No.	Topic	Description
A01	Accident tracker	An app records sudden movements of smartphones which indicate that a person fell down or had another accident and informs relatives
A02	Cardbox Headset Viewers	Smartphones put in a box create a headset viewer which allows simple augmentations of reality by combining images from the smartphone camera with additional data
A03	Crochet Style Community	An online platform enables crochet practitioners to exchange style suggestions, design patterns and further comments with each other and allow the purchase of wool and crochet tools
A04	Emotion-Responsive Interaction	Visual scans of faces are used to estimate the emotional state of a computer user to adapt the system operation to his or her current condition
A05	Gamified Learning	Different game designs are applied in the context of learning and training of physical and mental activities for education and health
A06	HiFi-Headphones Chair	A chair is equipped with mechanical devices which translate music played in headphones to simulate the resonance of physical objects to sound using different parametrizations
A07	Jewellery Configurator	An online configurator lets users pick from a large variety of design options for jewellery and provides further useful information about the products
A08	Personalized Advertising	Based on visual information about passengers in taxis and other vehicles, screens show them advertising for products which are likely to be interesting for them
A09	Product Information App	Customers in a retail store can use apps to identify products in a shelf and receive additional information about their function and usage
A10	Responsive Robot Toolkit	A construction toolkit allows the design of a robot out of flexible parts with different sensors which can be connected to external processing units
A11	Sensor Fitness Shirt	A sensorially enhanced shirt records health data from its users, controls their performance and suggests improved training routines
A12	Shoe Recommender System	Feet are scanned and compared to the measures of shoes. After adding further selection criteria, the systems suggests suitable footwear
A13	Sound-Based Vase Print	Visual representations of audio recordings from customers are used to generate an individually shaped vase with 3d printing
A14	Wardrobe Sharing Platform	Users share pictures and comments about the content of their wardrobes on an virtual discussion platform and offer clothes to others
A15	Wearable Device Design Kit	Standard design sets with electronic parts are enhanced with additional elements and back-end function for the creation of individual, flexible wearable devices

The case study presented here covers fifteen units of investigation, consisting of different projects which were purposively sampled to include a large diversity of subjects. Table 1 gives an overview of the subjects which were addressed.

Data collection

Excluding the attendees of special events like public talks or dedicated workshops, the average number of visitors per month to the facilities rose during the observation period from around 2.000 to more than 3.000. More than half of the visitors stayed longer than 10 min on the premises; more than one third stayed longer than 20 min, which gave them time to engage in deep interactions with personnel and company representatives. 22% of the visitors left the premises after less than five minutes. These short-term visits did not allow much interaction and provide little data for evaluation. However, many visitors who spent more time in the laboratory explained that they had come to the premises before to take a first look and then decided to return later, often with family and friends. To avoid distractions from the innovation activity, the personnel did not intrude any further into the privacy of the visitors, but only collected information which was freely given. This information indicated that the professional background of the visitors was quite diverse, including engineering expertise among some visitors and diverse competences and experiences in sales, marketing and business administration among others. For the most part, however, the visitors had no specific affinity to the innovation projects in the laboratory other than being potential beneficiaries.

During the months in which the projects were run in the laboratory, a group of researchers conducted structured observational studies of verbal and non-verbal behaviour. The observations gave insight into the topics which were discussed and the dynamics of the conversation between the participants. The researchers also documented the attention to physical objects and displays of information, usage of tools and behaviour in terms of looks that were exchanged, gestures and movements (Cooper and Schindler 2003). In addition, interviews were performed with the personnel on location about their impression of the practices in exchange with the information systems, contrasts and conflicts between different groups and their resolution (Guillemette 2003). Standardised techniques for coding and categorising data according to Krippendorff (2004) and Mayring (2000) were used to identify recurring patterns in the data, which were then set in relation with each other by triangulating findings across the source material.

Case analysis

Patterns of expression

It became clear from direct observation, interviews, as well as the original material provided by the participants that there was a large semiotic diversity in the approaches to the subjects. Distinctive patters could be recognised in at least four different aspects of verbal and non-verbal expression.

- Regarding the vocabulary, different groups of participants used specific, technical terminology which was not accessible to others and served in particular for highlighting functional characteristics of the displayed solutions. This included aspects of data processing, physical function or even medical function in terms of the effects on the human body.
- With respect to the disclosure of personal information, many participants remained very abstract and general. Others talked exclusively about their personal opinions or took over the personal perspective of friends and family members. Both groups had very little to say to each other.
- Non-verbal behaviour also allowed the distinction of people who touched everything and others who kept a strict distance. Both, however, allowed people to engage in lively discussions and share their thoughts and experiences.

- Another difference was recognised between discourses about the value of the projects for individual customers, for society, or for the progress of science and engineering. There was frequent exchange between people who found the projects technically interesting and others who recognised value in them for their daily life. Discourses on societal or wider economic issues, however, remained isolated within smaller groups of visitors.

Experiences of difference

Participants who did not use the same pattern of expression often had difficulties to relate, which became apparent in the little amount of direct engagement between them. Many had problems to follow the arguments of others and frequently misunderstood what they meant. The companies and their offerings were discussed extensively from contrasting perspectives, without convergence to a common view. The following examples illustrate this phenomenon.

(1) The product information app (A09) allowed customers in a shop to scan barcodes or price tags to gain further information about products, in particular when they required further assembly or were complicated to handle. Technical discussions focussed on the speed and reliability of the scan, search in the database, and proper display of the given information. From a commercial perspective, A09 was discussed as an additional service for customers, providing easy access to information from the producers. From the user perspective, the focus was set on the learning necessary to work with the system and to integrate it in daily routine. Statements included many personal judgements about the quality and appropriateness of the provided content. While some people were testing and discussing the technical function of the devices and their potential application in shops, others went on criticising the content and suggesting improvements. Some engaged in competitions assembling the product as described by the app, or found better ways of assembly on their own.

(2) The recommender system for shoes (A12) consisted of a scanning device to measure a customer's foot size and a database in which these measures were matched with suitable shoes. Discussions emerged about stores in the city which might be interested in the installation of such a system. A map was set up on which these stores were named, and there was a strong controversy on the role of internet retailers in this context. Some participants, however, were not interested in this issue at all, but instead fascinated by the three-dimensional image of their feet they saw. Some engaged in extensive conversations about orthopaedic problems with footwear and how the system let them address these problems independently from medical doctors who do not share information.

(3) The sound-based vase prints (A13) used acoustic recordings to 3d-print vases in individual shapes. Visitors considered them as highly personal gifts, to be offered at special occasions like weddings or birthdays. They started heated discussions about better and more flexible recording techniques and possibilities to reproduce the original sound from the printed object. Another argument emerged on the business model to commercialise the idea. Visitors described different kinds of occasions on which such gifts seems appropriate, or added further detail to the setting of the printing process and the delivery of the product. Many people also used the opportunity to test the system on location and experiment with recordings of different kinds of noises they produced or played on devices they brought in, focussing more on sculpturing the vases than considering the content of the recording.

Treatment of conflicts and contradictions

Overall, technical design, craftsmanship, efficiency and effectiveness of company offerings, their content and commercialisation, but also adoption and appropriation by users, usage scenarios and personal implications were addressed in separate discourses and activities. Company representatives considered this as a distinctive characteristic of the setting. All other forms of engagement with customers and other stakeholders they had used had been guided by more specific purposes and anticipated certain kinds of interactions and contributions. In absence of such guidelines and anticipations, the interaction was much broader and unveiled unexpected aspects of their offerings. At the same time, it turned out to be immensely difficult for companies to take advantage of these insights, because it was unclear how to relate them to business. Potential customers, on the other hand, worried that the company offerings could not effectively serve the purposes they had in mind.

In search for possibilities to make the different perspectives fit to each other, many ideas for new technical designs, commercialisation strategies and application scenarios came up, which provided a fertile breeding ground for reviews of current innovation strategies, as illustrated by the abovementioned examples. They inspired, among others, ideas for interactive information sharing solutions connected with product information scans, ideas for health-oriented foot scan systems, and various new business models and technical concepts in the context of 3d-printing from sound waves. These innovations only seemed pursuable if companies changed themselves regarding competences, structure and positioning on the market. Companies could learn from this experience how many different options for future development existed and identify new market fits.

Discussion

Corporate foresight and translation in open laboratories

The case study shows that activities in open laboratories exceed the scope of the knowledge flows. Open laboratories are not only means of resource appropriation which expand the range of opportunities to develop new products, services or business models. They also provide a space in which companies can gain insight into the fundamental differences between their own understanding of business operations and the understanding expressed by others. Companies experience difficulties of making sense of the views and opinions expressed by the visitors of the laboratory, who approach innovation from a completely different background than the companies themselves. Innovation management turns the focus on the outcome of this act, i.e. the input from the visitors which actually seems useful for the company; for corporate foresight, the act of translation itself is important, too, because it forces companies to review their relations to other stakeholders in commercial interactions and reflect on their own role in exchange with others.

Drawing on cultural theory, open laboratories can be described as heterotopias in which contrasting understandings of business and innovation become accessible. They open up a 'third space' which allows corporate foresight to face the problem of different attributions of meaning inside and outside the company. Companies can learn to appreciate outside contributors not only as additional resources in their own conceptions of value creation, but as stakeholders with alternate understandings of business. Cultural theory describes translation as an ongoing process in which the relation between different cultures is constantly re-negotiated. Ensuring that such re-negotiations take place can be considered as an essential aspect of corporate foresight which differentiates it from other corporate functions.

Theoretical and practical implications

This paper contributes to scientific research and managerial practice in various different ways. It introduces the concept of translation to research about open laboratories and corporate foresight, which drafts a new approach to interaction across organisational domains. This approach acknowledges the possibility of fundamental incompatibilities between the views of different stakeholders in commercial interactions and makes it possible to investigate their role of corporate foresight in symbiosis with innovation management. Furthermore, the paper presents evidence of translation processes in open laboratories. It shows how such translation processes add to a better awareness of a company's situation and its options for further development. The findings indicate that open laboratories can help companies to think out of the box and explore new directions for further business opportunities. In this respect, open laboratories can play an important role as a setting for corporate foresight exercises.

Due to the theoretical background for the study of open laboratories in this paper, many other aspects which might be relevant to corporate foresight have not been addressed. The paper has investigated how open laboratories can further enlarge the scope of open and interventional forms of corporate foresight. Open laboratories can also contribute to open foresight in different ways, for examples as sources to big data analytics regarding customer behaviour, expectations and perceptions of value. Empirically, the paper has focussed on qualitative insights gained from direct interaction between different groups of people present in the chosen location. It remains unclear to what extent the same results can be achieved in other types of open laboratories with less flexible settings. Future research will be necessary to explore other options to make use of open laboratories in corporate foresight, and to understand how they can be systematically developed.

Acknowledgements

The author wishes to express his gratitude for the support provided by everyone involved in the service innovation initiative of the Fraunhofer Institute for Integrated Circuits and the Innovation laboratory JOSEPHS in Nuremberg.

Disclosure statement

No potential conflict of interest was reported by the author.

References

Adner, R., and R. Kapoor. 2010. "Value Creation in Innovation Ecosystems: How the Structure of Technological Interdependence Affects Firm Performance in New Technology Generations." *Strategic Management Journal* 31 (3): 306–333.

Almirall, E., M. Lee, and J. Wareham. 2012. "Mapping Living Labs in the Landscape of Innovation Methodologies." *Technology Innovation Management Review* 2 (9): 12–18.

Anderson, P. D., B. H. Jorgensen, L. Lading, and B. Rasmussen. 2004. "Sensor Foresight—Technology and Market." *Technovation* 24 (4): 311–320.

Bassnett, S. 2013. *Translation*. London: Routledge.

Battistella, C., and A. F. De Toni. 2011. "A Methodology of Technological Foresight: A Proposal and Field Study." *Technological Forecasting and Social Change* 78 (6): 1029–1048.

Benjamin, W. 1968. "The Task of the Translator." In *Illuminations*, edited by H. Arendt, 69–82. New York: Schocken Books.

Benkler, Y. 2011. "The Unselfish Gene." *Harvard Business Review* 89 (7/8): 77–85.

Bhabha, H. K. 2004. *The Location of Culture*. London: Routledge.

Bergvall-Kåreborn, B., and A. Ståhlbröst. 2009. "Living Lab: An Open and Citizen-Centric Approach for Innovation." *International Journal of Innovation and Regional Development* 1 (4): 356–370.

Boeri, A., D. Longo, V. Gianfrate, and V. Lorenzo. 2017. "Resilient Communities. Social Infrastructures for Sustainable Growth of Urban Areas. A Case Study." *International Journal of Sustainable Development and Planning* 12 (2): 227–237.

Boxenbaum, E., and L. Rouleau. 2011. "New Knowledge Products as Bricolage: Metaphors and Scripts in Organizational Theory." *Academy of Management Review* 36 (2): 272–296.

Callon, M. 1986. "Some Elements of a Sociology of Translation." In *Power, Action and Belief: A New Sociology of Knowledge*, edited by J. Law, 196–233. London: Routledge and Kegan Paul.

Chesbrough, H. 2003. "The Logic of Open Innovation: Managing Intellectual Property." *California Management Review* 45 (3): 33–58.

Cooper, D. R., and P. S. Schindler. 2003. *Business Research Methods*. 8th ed. New York: McGraw-Hill.

Cuhls, K. 2003. "From Forecasting to Foresight Processes – New Participative Foresight Activities in Germany." *Journal of Forecasting* 22 (2-3): 93–111.

Cunha, M. P., J. V. Cunha, and S. R. Clegg. 2008. "Improvisational Bricolage: A Practice-Based Approach to Strategy and Foresight." In *Handbook of Research on Strategy and Foresight*, edited by L. A. Costanzo and R. B. MacKay, 182–199. Cheltenham: Elgar.

Daheim, C., and G. Uerz. 2008. "Corporate Foresight in Europe: From Trend Based Logics to Open Foresight." *Technology Analysis and Strategic Management* 20 (3): 321–336.

De Arias, A. R., S. D. Masi, D. Dorigo, F. A. Rojas, M. C. Vega, and M. Rolon. 2014. "Living Labs, Spaces for Open Innovation and Technology Transfer. An Alternative to the Solution of Social Problems in Paraguay." *Social Sciences* 3 (3): 74–79.

Dehaene, M., and L. De Cauter, eds. 2008. *Heterotopia and the City*. London: Routledge.

Denzin, N. K., and Y. S. Lincoln. 2005. "Introduction – The Discipline and Practice of Qualitative Research." In *The Sage Handbook of Qualitative Research*, 3rd ed., 1–32. Thousand Oaks: Sage.

Durst, C., M. Durst, T. Kolonko, A. Neef, and F. Greif. 2015. "A Holistic Approach to Strategic Foresight: A Foresight Support System for the German Federal Armed Forces." *Technological Forecasting and Social Change* 97: 91–104.

Eisenhardt, K. M. 1989. "Building Theories from Case Study Research." *Academy of Management Review* 14 (4): 532–550.

Fischer, G. 2013. "Learning, Social Creativity, and Cultures of Participation." In *Learning and Collective Creativity: Activity-Theoretical and Sociocultural Studies*, edited by A. Sannino and V. Ellis, 198–215. London: Routledge.

Foucault, M. 1971. *The Order of Things*. New York: Vintage Books.

Fritzsche, A. 2017. "Open Innovation and the Core of the Engineer's Domain." In *Philosophy and Engineering: Exploring Boundaries, Expanding Connections*, edited by D. P. Michelfelder, B. Newberry and Q. Zhu, 255–266. Amsterdam: Springer.

Gawer, A., and M. A. Cusumano. 2014. "Industry Platforms and Ecosystem Innovation." *Journal of Product Innovation Management* 31 (3): 417–433.

Genocchio, B. 1995. "Discourse, Discontinuity, Difference: the Question of Other Spaces." In *Postmodern Cities and Spaces*, edited by S. Watson and K. Gibson, 35–46. Oxford: Blackwell.

Gershenfeld, N. 2008. *Fab: The Coming Revolution on Your Desktop-From Personal Computers to Personal Fabrication*. New York: Basic Books.

Guillemette, F. 2003. *Handbook of Interview Research: Context and Method*. Thousand Oaks: Sage.

Halverson, E. R., and K. Sheridan. 2014. "The Maker Movement in Education." *Harvard Educational Review* 84 (4): 495–504.

Hatch, M. 2014. *The Maker Movement Manifesto: Rules for Innovation in the New World of Crafters, Hackers, and Tinkerers*. New York: MacGraw-Hill.

Huizingh, E. K. 2011. "Open Innovation: State of the Art and Future Perspectives." *Technovation* 31 (1): 2–9.

Hyysalo, S., C. Kohtala, P. Helminen, S. Mäkinen, V. Miettinen, and L. Muurinen. 2014. "Collaborative Futuring With and by Makers." *CoDesign* 10 (3-4): 209–228.

Jefferson, M. 2012. "Shell Scenarios: What Really Happened in the 1970s and What May be Learned for Current World Prospects." *Technological Forecasting and Social Change* 79 (1): 186–197.

Klag, M., and A. Langley. 2013. "Approaching the Conceptual Leap in Qualitative Research." *International Journal of Management Reviews* 15 (2): 149–166.

Krippendorff, K. 2004. *Content Analysis: An Introduction to Its Methodology*. Thousand Oaks: Sage.

Latour, B. 1990. "Technology is Society Made Durable." *The Sociological Review* 38 (S1): 103–131.

Leminen, S., M. Westerlund, and A. G. Nyström. 2012. "Living Labs as Open-Innovation Networks." *Technology Innovation Management Review* 2 (9): 6–11.

Levén, P., and J. Holmström. 2008. "Consumer Co-Creation and the Ecology of Innovation: A Living Lab Approach." IRIS 31, the 31st information systems research seminar in Scandinavia.

Liff, S. 2003. "Shaping e-Access in the Cybercafé: Networks, Boundaries and Heterotopian Innovation." *New Media and Society* 5 (3): 313–334.

Massa, L., and C. L. Tucci. 2013. "Business Model Innovation." *The Oxford Handbook of Innovation Management* 20 (18): 420–441.

Mayring, P. 2000. *Qualitative Content Analysis: Theoretical Foundation, Basic Procedures and Software Solution*. Klagenfurt: Gesis.

McPhee, C., M. Westerlund, and S. Leminen. 2012. "Editorial: Living Labs." *Technology Innovation Management Review* 2 (9): 3–5.

Miles, M. B., and A. M. Huberman. 1994. *Qualitative Data Analysis: An Expanded Sourcebook*. Thousand Oaks: Sage.

Möslein, K. M., and A. Fritzsche. 2017. "The Evolution of Strategic Options, Actors, Tools and Tensions in Open Innovation." In *Open Innovation Strategy and Communication for Innovation*, edited by N. Pfeffermann and J. Gould, 61–76. Berlin: Springer.

O'Hern, M., and A. Rindfleisch. 2008. "Customer Co-Creation: A Typology and Research Agenda." *Review of Marketing Research* 6: 94–106.

Ramirez, R., L. Roodhart, and W. Manders. 2011. "How Shell's Domains Link Innovation and Strategy." *Long Range Planning* 44 (4): 250–270.

Ratcliffe, J. S. 2005. "Challenges for Corporate Foresight: Towards Strategic Prospective Through Scenario Thinking." *Foresight* 8 (1): 39–54.

Rhisiart, M., R. Miller, and S. Brooks. 2015. "Learning to Use the Future: Developing Foresight Capabilities Through Scenario Processes." *Technological Forecasting and Social Change* 101: 124–133.

Richardson, M., S. Elliott, and B. Haylock. 2013. "This Home is a Factory: Implications of the Maker Movement on Urban Environments." *Craft + Design Enquiry* 5: 1–3.

Rohrbeck, R., C. Battistella, and E. Huizingh. 2015. "Corporate Foresight: An Emerging Field with a Rich Tradition." *Technological Forecasting and Social Change* 101: 1–9.

Rohrbeck, R., and H. G. Gemünden. 2011. "Corporate Foresight: Its Three Roles in Enhancing the Innovation Capacity of a Firm." *Technological Forecasting and Social Change* 78 (2): 231–243.

Rohrbeck, R., F. Steinhoff, and F. Perder. 2010. "Sourcing Innovation from Your Customer: How Multinational Enterprises Use Web Platforms for Virtual Customer Integration." *Technology Analysis & Strategic Management* 22 (2): 117–131.

Rosa, H. 2010. *Alienation and Acceleration: Towards a Critical Theory of Late-Modern Temporality*. Natchitoches: NSU Press.

Roth, A., A. Fritzsche, J. Jonas, F. Danzinger, and K. M. Möslein. 2015. "Spaces for Value Co-Creation: The Case of "JOSEPHS® – The Service Manufactory"." EURAM conference, Warsaw.

Ruff, F. 2015. "The Advanced Role of Corporate Foresight in Innovation and Strategic Management — Reflections on Practical Experiences from the Automotive Industry." *Technological Forecasting and Social Change* 101: 37–48.

Rutherford, J. 1990. *Identity: Community, Culture, Difference Identity*. London: Lawrence and Wishart.

Saldanha, A. 2008. "Heterotopia and Structuralism." *Environment and Planning A* 40: 2080–2096.

Toupin, S. 2014. "Feminist Hackerspaces: The Synthesis of Feminist and Hacker Cultures." *The Journal of Peer Production*. 5: 1–9.

Troxler, P. 2015. *Beyond Consenting Nerds: Lateral Design Patterns for new Manufacturing*. Rotterdam: Hogeschool Rotterdam Uitgeverij.

Troxler, P. 2016. "Fabrication Laboratories (Fab Labs)." In *The Decentralized and Networked Future of Value Creation*, edited by J.-P. Ferdinand, U. Petschow, and S. Dickel, 109–127. Heidelberg: Springer.

von der Gracht, H., C. R. Vennemann, and I. L. Darkow. 2010. "Corporate Foresight and Innovation Management: A Portfolio-Approach in Evaluating Organizational Development." *Futures* 42 (4): 380–393.

von Hippel, E. 2009. "Democratizing Innovation: The Evolving Phenomenon of User Innovation." *International Journal of Innovation Science* 1 (1): 29–40.

Voytenko, Y., K. McCormick, J. Evans, and G. Schliwa. 2016. "Urban Living Labs for Sustainability and Low Carbon Cities in Europe: Towards a Research Agenda." *Journal of Cleaner Production* 123: 45–54.

Warf, B., and S. Arias. 2008. *The Spatial Turn: Interdisciplinary Perspectives*. London: Routledge.

Westerlund, M., and S. Leminen. 2011. "Managing the challenges of Becoming an Open Innovation Company: Experiences from Living Labs." *Technology Innovation Management Review* 1 (1): 19–25.

Wooffitt, R. 2005. *Conversation Analysis and Discourse Analysis: A Comparative and Critical Introduction*. Thousand Oaks: Sage.

Yin, R. K. 2009. *Case Study Research: Design and Methods*. Thousand Oaks: Sage.

Competitive intelligence practices of European firms

Jonathan Calof, Rubén Arcos and Nisha Sewdass

ABSTRACT

This article presents the results of a survey of competitive intelligence (CI) practices in European firms. In comparing the results to a similar 2006 global study and a 2006 European CI study, it appears that the breadth of applications for CI has grown well beyond competitors to include customer related intelligence, technology, market, etc. Innovation is driving much intelligence activity, in particular research and development (R&D) and new product development decisions. CI is more formalised now in European firms than it was in 2006. The study also found similarities between corporate foresight and CI in terms of objectives (development and maintenance of competitive advantage, help with decision making) and analytical techniques with scenarios being among the more frequently used analytical techniques along with STEEP and other environmental analysis in both corporate foresight studies and the CI study.

Introduction

As business environments become increasingly volatile, 'countries and companies rely on early detection of environmental changes so that they may respond' (Kahaner 1996, 23). Since firms need time to adapt to changing environments, they need to develop the ability to anticipate changes and determine consequences of alternative responses to those changes. In this volatile and competitive environment, competitive intelligence (CI) has emerged and grown into a discipline to help organizations adapt to environmental change (Sewdass 2012) and deal with industry disruption (Vriens and Soilen 2014). CI improves decision-making as it provides a framework for including integrated environmental analysis in decision-making and as such, has the potential to help a company meet or exceed its objectives and business goals (Hitt, Ireland, and Hoskisson 2000). Qiu (2008) found that CI helped companies develop and maintain their competitive advantage. M-Brain (2015) in their study on CI noted that most respondents had benefited from intelligence and had a high return on their intelligence investment. CI activities appear to have grown in the past 10 years as measured by companies with formal intelligence structures and personnel, membership in professional CI associations (Calof 2014; Erickson and Rothberg 2012) in fact, 2016 du Plessis et al wrote that 'CI is one of the fastest growing fields in the business world'(page 3).

Despite the growth in CI and its positive impact, it has been10 years since a comprehensive look was done on CI practices around the world (Fehringer, Hohhof, and Johnson 2006) and since reporting on European wide CI practices (Wright and Calof 2006). Therefore, the intent of this paper is to identify current European CI practices and compare these to those from the 2006 studies. Finally, given the similarities between corporate foresight and CI (Calof and Smith 2009; Calof, Richards, and Smith 2015; Hammoud and Nash 2014), a comparison is made of the results of this study with corporate foresight study results.

Literature review

Competitive intelligence

In 2015, Du Toit examined 338 articles about CI looking at trends in the literature between 1994 and 2014. The article provides examples of the evolution of the definition of CI and includes a definition based on meta-analysis of CI definitions

> a process or practice that produces and disseminates actionable intelligence by planning, ethically and legally collecting, processing and analysing information from and about the internal and external or competitive environment in order to help decision-makers in decision-making and to provide a competitive advantage to the enterprise (page 15).

The Competitive Intelligence Ning (a discussion forum for competitive intelligence practitioners), defines CI as 'the interpretation of signals from the environment for an organisation's decision makers to understand and anticipate industry change' (Competitiveintelligence.ning.com 2016). Du Toit (2013, 30) defined the purpose of CI as 'a strategic tool to facilitate the identification of potential opportunities and threats.' Plessis and Gulwa (2016) also saw CI as being a strategic tool as did Jenster and Soilen (2013). These and others see CI as a process to assist in making key decisions.

Researchers have looked at kinds of decisions where CI can be applied. Amara, Soilen, and Vriens (2012, 30) wrote 'CI enables senior managers in companies of all sizes to make informed decisions about everything from marketing, R&D, and investing tactics to long-term business strategies.' Moreover, CI is considered a value-added concept that outperforms the top of business development, market research and strategic planning. Vriens and Soilen (2014) wrote about intelligence being used to see whether industry disruption is going to occur. They termed this disruptive intelligence.

Recently, a broader definition of CI has emerged – integrated intelligence. This is defined as

> the robust integration of insights from "intelligence pools" that are identified across the business environment and in collaboration with other functional areas and disciplines that are synthesized to gain a comprehensive picture of a market in its current state and in its probable future state. The resulting outcome of integrated intelligence efforts is critical decisions influencing and supporting recommendations required to drive and gain a competitive advantage for an organization (Bulger 2016, 63).

This view of CI was used for the study.

Competitive intelligence and foresight

In defining foresight, FOR-LEARN included intelligence: 'Foresight is a systematic, participatory, future-intelligence-gathering … .' (FOR-LEARN 2017). This view has intelligence as being part of the foresight process or at least supporting it. This perspective is shared in Canongia, Antunes, and Freitas Pereira (2004). They looked at applying CI tools to help develop technology foresight, in particular, to identify international trends. Hammoud and Nash (2014, 41) have a similar view but in their case, it's foresight supporting CI 'in the related field of competitive intelligence, researchers recommend future-oriented methods such as the use of scenario analysis.' Similarly, Olsmats and Kaivo-oja (2014, 38) in their study of European packaging industry foresight wrote that the foresight approach 'provides new competitive intelligence for companies and corporations.'

Others have pointed to the similarity and relatedness of foresight and CI much in the same way as the earlier quote from Hammoud and Nash (2014). Calof and Smith (2009, 31) wrote 'Foresight and CI are two fields that seek to address future oriented environmental scanning.' Similarly, Neugarten (2006, 894) wrote 'The practice of both CI and foresight attempt to prevent strategic surprise by noticing and attending to signals earlier rather than later.' Both CI and corporate foresight are seen as future-oriented environmental scanning that attempt to prevent strategic surprises. Even the core intent of both domains is somewhat similar. For example, Arrigo (2016, 51) suggests that the main output of the CI process should be the ability to make good forward-looking decisions in order to be the market leader.

Even when looking at why corporate foresight and CI is done, similarities are evident. In their European corporate foresight study, Daheim and Uerz (2008) found that corporate foresight was used for achieving competitive advantage and supporting strategic decision-making. Furthermore, Hammoud and Nash's (2014) study of corporate foresight reported that almost half of the responses identified innovation or competitive advantage as the primary reasons for corporate foresight. Rohrbeck, Battistella, and Huizingh (2015, 2) suggest that corporate foresight

> permits an organization to lay the foundation for future competitive advantage. It entails identifying, observing and interpreting factors that induce change, determining possible organization-specific implications, and triggering appropriate organizational responses. Furthermore, CF involves multiple stakeholders and creates value through providing access to critical resources ahead of competition, thereby preparing the organization for change, and permitting the organization to steer proactively towards a desired future'.

Similarly, Fehringer, Hohhof, and Johnson (2006) reported that CI was being used to support strategic decision-making. Arrigo (2016, 51) found that CI helps to strengthen the strategic planning process by facilitating the choice of the competitive strategies to implement.

While the fields have similarities, they have differences. For example, Calof and Smith (2009) noted that each had different temporal focus with CI being shorter time oriented than foresight. As well, the authors noted that the orientation of each was different, with CI starting with the planned strategy and studying its consistency with the environment while foresight tended to start with the external environmental look (e.g. scenario's) and then looking at the impact of that assessment on the organizations strategy. The authors referred to this as intelligence being inside out and foresight being outside in.

Given the similarities between CI and corporate foresight a comparison of the results of this study will be made with the results of studies on corporate foresight, including studies on European corporate foresight.

Competitive intelligence in Europe

Most of the European CI practice studies done are now 10 years or older and this study hopes to get a more current reflection of the practice in European firms.

One of the early studies on European CI was Badr, Wright, and Pickton (2004). Central to the objectives of the study were issues of the current state of CI in European companies, whether a relationship between CI and Marketing Strategy Formulation (MSF) existed at all and how CI contributed to the marketing strategy process. The findings revealed the following:

- Firms recognised the importance of CI and as such, were more likely to value its contribution.
- Over half of the respondents used the term 'CI'.
- Many European companies used their CI activities in strategic planning which gave both long-term and short-term benefit.
- The attitude of senior managers about CI was largely unknown by most respondents and some indicated that the lack of support from senior managers was the main barrier to the development of CI in their company.
- The tools/systems used to acquire, access, store, and share CI were 'database' and 'secure intranet'. However, they found that although companies were investing in such tools, there were still many problems facing CI managers concerning this issue. Even though firms had such systems in place,

many senior managers still relied heavily on the CI manager to provide even the simplest intelligence items.

- CI managers mainly use Strengths Weaknesses Opportunities Threats (SWOT) analysis and Competitor Profiling. They were less likely to use STEP/PEST analysis, war gaming/role-playing and win/loss analysis. The main reason for only using SWOT analysis was owing to their limited knowledge of other techniques.
- The result of this research indicated that the majority of respondents used 'action taken' and 'market share improvements' as being the main type of measure used to assess the effectiveness of CI.
- The respondents in this study indicated that while they understood the importance, benefits and need for CI in the MSF process, their main problem was in 'how' to integrate CI into MSF.
- CI managers believed that CI often contributed to strategic analyses and strategic decision making. They also believed that CI 'sometimes' contributed to setting marketing objectives, implementation, and control.

Another study by Badr, Madden, and Wright (2006) investigated CI practices in the European pharmaceutical industry in relation to the contribution CI made to the strategic decision making (SDM) process. It was hoped that the outcome would provide insight into how CI was used in this industry as well as identifying any issues or problems, which surround the current practice. The findings revealed that many pharmaceutical companies in Europe had recognised the importance of CI as an essential tool in understanding the external business environment and in providing the intelligence needed for decision-making. It was also found that although the CI activities in respondent firms were reasonably well established and appreciated, there still remained a challenge for the industry as it was not taking full advantage of CI (Badr, Madden, and Wright 2006, 16). These and other European studies are summarised in Wright and Calof (2006).

There have been individual European country. For example, the Journal of Competitive Intelligence in Management had a special issue devoted to country-specific intelligence in 2004 which included Wright, Badr, and Weiss (2004) writing about CI in the United Kingdom, as well as articles looking at CI in Finland and in Israel. Competitive Intelligence Review had articles about competitive intelligence in France in the1990's. However, most of these articles are also well over 10 years old with the exception of Barnea who looked at CI in Israel in 2016. There have been a few global studies of intelligence with regional write ups, most notably from Global Intelligence Alliance and M-Brain. They have conducted global studies of CI since 2005. However, these studies have focused on large global companies and the studies have not provided comprehensive reporting on CI practices but only isolated elements.

Methodology

The results reported in this paper constitute the first phase of a global study on CI practices. The first phase was development of the CI survey and testing it on European companies. Europe was chosen as the timing of the study allowed for the research team the opportunity to gather data and present the study at the SCIP Europe Conference.

Survey design

The main objective of the study was to look at CI practices around the world to identify the current practices. To enable comparisons between the results of this study and these past ones mentioned earlier, the researchers used as the Fehringer et al. questionnaire as the base for this study (Wright and Calof 2006 used similar questions). The Fehringer et al. study had 28 questions. These questions were developed following an extensive literature review as well as asking leading CI practitioners and academics to review the questionnaire (Fehringer, Hohhof, and Johnson 2006, 7). For this study, additional questions were added based on developments in CI practice as reported either in the academic literature or the professional literature. The questionnaire (revised from the Fehringer study) was then sent to five leading CI academics and practitioners for comment and validation. Based on these comments, additional changes were made resulting in a questionnaire with 42 questions which included 10 personal identification questions (how many years respondent has practiced CI, percent of time that they conduct CI activities), two company classification questions (size and industry) and the rest on respondent's competitive intelligence activities. More details on the specific questions are provided in the results section.

To ensure that respondents answered the questions in the context of the integrated intelligence concept mentioned in the literature review section, the survey cover letter provided respondents with the following CI definition:

> CI is a necessary, ethical business discipline and/or skillset for decision making based on understanding the competitive environment in order to drive to competitive advantage in a marketplace. Any organizational employee who is gathering insights on the external environment (competitors, customers, suppliers, technology, etc.) in order to make decisions is practicing a form of CI. CI validates decision making by introducing a disciplined system not only to gather information, but also to perform analysis and disseminate findings about the external environment tailored with the intent to drive competitive advantage for their organization.
>
> In referring to the above definition of CI, we recognize that there are no consistencies in the terms used by organizations and the frequently used terms are environmental scanning, market intelligence, business intelligence, foresight, CI and so forth.

Survey delivery

An online survey supported by Strategic and Competitive Intelligence Professionals (SCIP) was conducted. Online surveys are the predominant procedure for survey data collection (Callegaro, Manfreda, and Vehovar 2015). Consistently with the survey methodology literature, surveys are considered here as a systematic method for collecting information from a sample of individuals for the purpose of constructing statistics of the attributes of the population of which those individuals are members (Groves et al. 2009, 2).

Data collection was conducted late summer-fall 2015, starting with promotion of the survey in September in the SCIP Newsletter. This newsletter is sent to all SCIP members. In addition, European SCIP chapter chairs were asked to send an email to those on their chapter contact list reminding them about the survey and asking them to complete it. Two notifications were sent from SCIP via the newsletters and SCIP European chapter chairs were emailed on three separate occasions. It was thought that by receiving a notification about the survey from the association (SCIP) and also email from the local chapter chair response rates would be higher. Additionally, a hard copy of the survey questionnaire was made available at the SCIP European Summit (November) at a session were the preliminary results were presented.

Sample frame and response rate

The targeted population for the European CI practices survey was European members of SCIP and others who were on the European SCIP chapter mailing lists. Both the Fehringer, Hohhof, and Johnson (2006) and Wright and Calof (2006) write up of the European studies used SCIP members as their primary sample frame as well.

A total of 156 European responses were received. This compares to 62 European respondents from the 2006 study of Fehringer et al. To help establish response rates, the survey asked respondents if they were SCIP members. By then dividing the number of respondents who identified themselves as SCIP members by the total number of SCIP members in Europe (according the SCIP member database), it was possible to establish a response rate of 25.6% for European SCIP members. It is not possible to estimate the response rate of non-SCIP members since there is no data on how many additional individuals were informed on the online survey beyond the SCIP membership lists or may have been emailed about the survey from those that received the SCIP information.

Non-response is a prominent problem in survey research (Peytchev 2012). In the case of organisational research, survey response rates are declining (Fulton 2016; Rogelberg and Stanton 2007) which can threaten the external validity of conclusions drawn. Thus, being aware of survey fatigue and declining response rates (Fulton 2016) a response rate of 25.6% was considered good.

Respondents came from a broad range of industries. The most frequently mentioned were CI or consulting (19%), telecommunications (13%), pharmaceutical/biotechnology or healthcare (10%), manufacturing (10%), consumer goods or services (9%), software (8%) and financial services or insurance (6%). In addition, 62% of respondents came from companies with over 1000 employees with 12% having less than 10 employees. The respondents had on average 7.5 years of experience in CI and spent, on average, more than 50% of their time on CI activities with 32% saying that they spent almost all their time involved with CI.

Empirical results

The objective of the study and this article was to examine European CI practices. In this section, these results are presented and compared where possible with past European studies and the Fehringer, Hohhof, and Johnson (2006) study. The latter results were aggregated at the global level but do provide insight to help examine whether intelligence practices have changed in the past 10 years. In addition, given the similarities between CI and corporate foresight as described in the literature review section comparisons will also be made with Daheim and Uerz's 2008 study of corporate foresight in Europe and Hammoud and Nash's 2014 study on corporations use of foresight.

Intelligence structure and process

Respondents were provided with a list of five different types of structures for intelligence (including having no formal structure) and eight different titles for the unit (if it existed). Tables 1 and 2 present these results. The two most dominant CI structures were centralised units (one in the organisation – 42%) followed by a mixed approach where there was a centralised unit and also additional intelligence activities conducted throughout the organisation (34%). The previous European study reported in Wright and Calof's (2006) article noted that 18% had one full-time CI person and of those, 23% had a separate CI department. It appears that over the 10 years there has been an increase in formal CI activities.

Regarding the name of the department responsible for CI – CI and business intelligence were #1 and #2 respectively, but of note was that 10% of respondents indicated multiple departments responsible for CI. Frequently mentioned were both a CI unit and a business/customer insight unit. Wright and Calof (2006) and Wright, Badr, and Weiss (2004) reported that the most frequently mentioned department responsible for CI was marketing. This again provides evidence that since 2006 there has been an increase in the development of CI in Europe as a separate function.

When asked how many full time employees they had supporting CI, companies responded that they had on average 2–4 employees (31%), 28% having 1 or less, 17% had 5%–9% and 24% had

Table 1. Structure of the intelligence function.

Type of structure	Percent
Centralised: one CI function serves all or most of the organisation	42
Decentralized: each department or functional line of business does its own competitive intelligence	10
Hub and spoke organisation but centrally driven	1
Informal: no structured CI function at any level, CI tasks occur only as needed and may or may not be labelled as 'CI'	13
Mixed: some activities are centralised, others are done independently by Individual departments or functional lines of business	34

Table 2. Name of department responsible for competitive intelligence.

Name of unit	respondents (%)
Competitive intelligence	27
Marketing intelligence	15
Marketing/market research	12
Strategic planning	11
Multiple units	10
Business intelligence	7
Market insight	5
Competitor insight	3
Other names	10

greater than 10. The Wright and Calof 2006 article noted that 18% had one full-time CI person with the rest of the companies indicating that it was a part-time function. This again shows increase in development in CI in Europe since 2006.

Relating these results to CF, the Daheim and Uerz's (2008) study on European corporate foresight noted that foresight processes were run by specialised departments (in more than half the companies) with titles such as market intelligence and one-half had at least one staff member devoted to foresight. The results are very similar to those of this study in terms of the unit name but this study had a higher percentage of companies with full-time resources devoted to intelligence.

Competitive intelligence procedures

The 2006 study examined the extent to which CI had been formalised within responding organizations. This was operationalised using four questions also used in this study. 'Does your organization have:' 'Formal written down competitive intelligence strategy and procedures' (44% and 42% responded yes), specific CI ethical guidelines (48% – responded yes) and a manager function with CI responsibilities (70% responded yes). These responses are very similar to the 2006 study with the exception of managers with CI responsibility which was 50% in the 2006 study and 70% in this study.

The 2006 study also examined the extent of employee awareness of and involvement in CI. These question was also asked in this study. All respondents reported that there were employees in their organisation that were aware that CI existed and 96% indicated that employees had some involvement in CI activities (Table 3). The results are similar both to those reported in the Fehringer et al. study (2006) and the Wright and Calof's article (2006) and point to a high percentage of company employees being both aware of and involved in CI activities.

Intelligence focus and planning

12% of respondents' CI time involved planning and focusing CI projects. For this study, focus referred to both the clients that intelligence served and the topics or areas that intelligence examined. Respondents were asked the extent to which board of directors members, C level management and other management were clients of CI using a 5 point Likert scale -0, none to 4 all. Respondents

Table 3. Company-wide knowledge and involvement in CI.

	Organizations employees that know CI exists (%)	Organizations employees that participate in CI activities (%)
None	0	4
Few	14	24
Some	34	53
Most	37	12
All	15	7

indicated that each was on average between some and most, 2.2 for board of directors, 2.8 for C-level and 2.4 average for management below C level.

Regarding the focus of intelligence, respondents were asked about the business decisions supported by CI (Table 4), the targets of CI (the extent to which intelligence is focused on competitors versus customers versus government, etc. – Table 5), and the type of intelligence products developed (for example market entry reports, benchmarking – Table 6). Innovation related intelligence topics were frequently mentioned including R&D decisions and new product development. Compared to the 2006 Fehringer et al.'s study, it appears that the breadth of use of CI was higher. Most categories in Tables 4 and 6 was higher than in 2006. Also of note was that competitors as the sole or primary focus of CI declined. For example, in reporting on the European CI study, Wright and Calof (2006) revealed that the #1 reason for doing CI was to deal with competitors' activities (97%) versus this study that had under 50% of intelligence time focused on competitors. Increased breadth in CI focus are consistent with the integrated intelligence concept.

A final aspect of focus used in the study lies with the temporal orientation of the intelligence process, that is, how forward looking the intelligence was. Respondents indicated that 51% of intelligence projects looked forward less than one year, 37% looked forward 1–5 years and 12% looked forward more than 5 years. This is very similar to Hammoud and Nash's 2014 corporate foresight study where they reported on interviews that talked about two-time orientations with corporate

Table 4. Business decisions supported by CI in the organisation from 0 (never) to 3 (frequently)

	Average	responding frequently (%)
Corporate or business strategy	2.4	48
Sales or business development	2.2	40
Market entry decisions	2.2	48
Product development	2.1	39
Research or technology development	1.9	27
Mergers and acquisitions, due diligence, joint venture assessment	1.8	35
Reputation management/ communications/Public relations	1.5	17
Regulatory or legal	1.4	37

Table 5. Targets of intelligence efforts.

Target	of time focused on the target (%)
Competitors	48
Customers	21
Government	7
Suppliers	6
Partners	6
Universities	3
Professional associations	4
Other research institutions	5

Table 6. Competitive Intelligence products (0 never to 3 frequently).

	Average this study	responding frequently (%)
Market industry/report/ analysis	2.4	56
Benchmarking	2.4	55
Company profiles	2.3	47
Early warning alert	1.9	30
Technology assessments	1.7	23
Economic analysis	1.7	20
Customer profiles	1.6	18
Executive profiles	1.5	16
Political analysis	1.2	13
Supplier profiles	1.1	5

foresight – short-time horizon projects of less than two years (consumer research) and longer time horizons which were described as up to 4 years. Daheim and Uerz (2008) also write about corporate foresight time horizon as being shorter than public sector foresight timeframe and it appears that the time horizon of European CI projects are similar to those found in these corporate foresight studies.

Information sources used

A quarter (25%) of respondents' intelligence time was spent collecting information. Secondary information sources largely online such as the internet and commercial databases were the top sources of information used in the intelligence effort both in terms of average score and the percent of respondents reporting that the source was very important. In terms of primary sources, customers, company employees and industry experts were most frequently cited (Table 7). While this is similar to the 2006 Fehringer et al.'s study and Wright and Calof's (2006) article, internet has increased in importance as a source of information for CI and social media has also increased in importance. Trade shows were seen as being considerably more important in this study (29% saying very important) compared to Wright and Calof (2006) which reported the overall use of trade shows at 11%.

Analytical methods

About 28% of respondent intelligence time involved analysis. They were presented with 47 different analytical techniques found in the CI literature. The number is almost double those in the 2006 study reflecting increasing development in the field around analysis. The list was developed from those techniques assessed in the Fleisher and Bensoussan analysis books (2003, 2012). Eighty two percent responded that they used some kind of analytical model to develop intelligence. Most frequently used were competitor analysis (44%), SWOT (38%), benchmarking (36%), competitive positioning analysis (26%), industry analysis (25%), customer segmentation (23%), scenario (23%), and patent (21%) (See Table 8). Respondents used on average 10.2 analytical techniques. Regarding technology, 31% used social media analytics and 35% used business analytics.

While the top two analytical techniques (SWOT and competitor analysis) were the same as both the Fehringer et al. and Wright and Calof study's the breadth of techniques used was greater in this study. The results are also very similar to the Hammoud and Nash (2014) corporate foresight study which found scenario planning being among the most popular techniques, used by 18% of respondents (23% for this study) and environmental scanning at 10% with STEEP being used frequently. In their study on European corporate foresight practices, Daheim and Uerz (2008) reported trend analysis as the #1 technique used, followed by media publications analysis and scenarios. Scenario analysis was among the top techniques used both in this study and the corporate foresight

Table 7. Sources of information for CI (0 not important at all to 4 very important).

Information source	Score (0–4)	responding very important (%)
Internet websites (free)	3.5	67
Publications (print/online)	3.4	60
Commercial databases (fee)	3.1	52
Customers	3.0	40
Company employees	2.9	39
Internal databases	2.9	35
Industry experts	2.9	32
Trade show's/conferences	2.7	29
Social media	2.2	25
Suppliers	2.2	17
Association employees	1.7	6
Government employees	1.5	8

Table 8. Analytical methods used.

Analytical technique used	responding yes (%)	Analytical technique used	responding yes (%)
Analysis of competing hypothesis	5	Issue analysis	5
Analytics	9	Linchpin analysis	2
Benchmarking	36	Macro – environmental STEEP analysis	10
Blind spot analysis	9	Management profiling	11
BCG growth/share portfolio matrix	19	McKinseys 7s analysis	7
Business model analysis	16	Patent analysis	21
Competitive positioning analysis	26	Product life cycle	14
Competitor analysis	44	Product line analysis	6
Critical success factor analysis	8	S-curve (technology life cycle analysis)	8
Customer segmentation analysis	23	SERVO analysis	1
Customer value analysis	14	Scenario analysis	23
Data visualisation	16	Shadowing	2
Driving forces analysis	15	Stakeholder analysis	14
Experience curve analysis	2	Strategic group analysis	8
Financial ratio and statement analysis	15	Strategic relationship analysis	5
Financial analysis and valuation	19	Supply chain management analysis	7
Functional capability and resource analysis	8	Sustainable growth rate	5
GAP analysis	14	SWOT analysis	38
General Electric business screen matrix	3	Technology forecasting	10
Growth vector analysis	2	Timeline/event analysis	14
Historiographical analysis	6	Value chain analysis	14
Indications and warning analysis	8	War gaming	12
Industry analysis	25	Win/loss analysis	14
Industry fusion analysis	3		

studies and both sets of studies have noted that industry/environmental analysis also being among the most frequently used techniques.

Communication of intelligence findings

About 18% of respondent intelligence time was spent communicating intelligence findings. Email and presentations/staff briefings were the most frequently used methods for communicating intelligence findings (Table 9) with all respondents using multiple methods for communicating their findings. Wright and Calof's (2006) article reported that most popular methods for communicating intelligence were central databases (75%). Email and presentations were also frequently used, but at a significantly lower level (47% and 36%) than those in this study.

Assessing CI effectiveness

Managing CI projects including assessing their effectiveness took 17% of respondents CI time (6% in assessing effectiveness alone). The most frequently used method for assessing CI effectiveness was customer satisfaction followed by decisions made/supported. This was consistent with the 2006 Fehringer et al.'s study. Having no effectiveness measures, was among the most frequent responses both in this and Fehringer's study (Table 10).

Table 9. Communicating CI Results (0 never to 3 frequently).

	Average response	responding frequently (%)
E- mails	2.6	67
Presentations/staff briefings	2.3	59
Central database	1.9	47
Newsletters	1.8	42
Printed alerts or reports	1.8	41
Personal delivery	1.8	30
Teleconference	1.7	24
Company intranet	1.6	42
Warning alerts	1.5	22

Table 10. Assessing CI Effectiveness.

	of respondents using the metric (%)
Customer satisfaction	30
Decisions made/supported	26
New products or services developed	19
CI productivity/output	16
Strategies enhanced	16
New or increased revenue	14
We have no effectiveness measure	13
New products or services	12
Cost savings or avoidance	10
Financial goals met	10
Return on investment	10
Profit savings	9
Time savings	8

Conclusions and areas for future research.

This paper cited three objectives:

(1) Identify European CI practices;
(2) Compare the results of the current study to 2006 studies to see if there had been changes in CI practice; and
(3) Compare the results to corporate foresight practice studies to see whether there was a commonality between CI and corporate foresight practice.

156 European responses, 94 more than the 2006 study and a 25.6% rate from SCIP members, provided a good base of information to examine European CI practices. Structurally, 87% of all responding organizations had some form of formal CI structure and many organizations had multiple intelligence or intelligence type functions in their organisation. In only 4% of the responding companies were employees of the company not involved with intelligence activities. In terms of intelligence focus, intelligence was used by the responding companies for many different types of both strategic and tactical business decisions with competitors, customers and innovation being among the most frequent intelligence focuses. Compared to the 2006 study results, more European firms appear to have formalised intelligence functions and are using these intelligence functions for a broader range of intelligence objectives (range of projects and targets).

Regarding the collection side of intelligence, secondary sources, including social media have grown in importance since the 2006 studies but so to have trade shows for the intelligence process. Respondents used on average 10.2 analytical techniques. The role and importance of analysis has grown within intelligence but the more popular techniques, SWOT, competitors, benchmarking, competitive positioning, industry analysis and scenario analysis continue to be the most used techniques. Industry/environmental techniques and scenario analysis have also been found to be among the more often used corporate foresight techniques.

Email and presentations were the most used approaches used for communication intelligence with most respondents using multiple communications approaches. This was similar to the 2006 findings. Finally, assessing intelligence effectiveness, an area of increasing importance in CI was noted as important in this study with only 13% of respondents to this question not having any effectiveness measures. Most used measures were customer satisfaction and decisions made due to the intelligence.

In conclusion, the above summarises the key findings in terms of current CI practices of European firms. In comparing these to the results of the 2006 European and global studies, it appears that intelligence is more formalised and widespread now than it was in 2006. Given that the studies used a similar base of respondents (SCIP members) and base of questions, this conclusion has some merit. However, without the actual data from these studies, it is impossible to conclude whether the results are in fact statistically significant. One can only conclude that it appears it is more formal. The magnitude of some of the differences between 2006 studies and this one suggests that it is probable that the results would be statistically significant – for example, 18% of respondents having a formal CI resource in the 2006 study versus 83% in this study. Finally, in comparing the results of this study to prior corporate foresight studies, many similarities were found including objectives (supporting decision making and competitive advantage), temporal orientation (how forward looking the projects are), some of the analytical techniques used and in some cases the name of the unit conducting the corporate foresight/CI. This supports the notion that corporate foresight and CI are related to each other. To conclude this definitively, future research should compare results to a broader base of corporate foresight studies using statistical tests.

Future research could also look at whether these practices have an impact on firm performance. As well, future studies should look at whether the European companies practices cluster into different groups and the relationship between the cluster and performance. For example, Global Intelligence based on respondent answers on questions dealing with intelligence scope, intelligence process, intelligence deliverables, intelligence tools, intelligence organisation and intelligence culture classify respondents on a five-point scale ranging from informal intelligence to world class intelligence.

Acknowledgements

The authors thank SCIP (strategic and competitive intelligence professionals) for their support of this research. Dr. Jonathan Calof wants to acknowledge that his portion of the article was prepared within the framework of the Basic Research Program at the National Research University Higher School of Economics (HSE) and supported within the framework of the subsidy granted to the HSE by the Government of the Russian Federation for the implementation of the Global Competitiveness Program. Dr. Rubén Arcos wants to thank the Centre for Intelligence Services and Democratic Systems at Rey Juan Carlos University in Spain for the support during the preparation of his portion of the article.

Disclosure statement

No potential conflict of interest was reported by the authors.

ORCID

Jonathan Calof http://orcid.org/0000-0002-2146-1506
Rubén Arcos http://orcid.org/0000-0002-9665-5874

References

Amara, Yasmina, Klaus Solberg Soilen, and Dirk Vriens. 2012. "Using the SSAV Model to Evaluate Business Intelligence Software." *Journal of Intelligence Studies in Business* 2 (3): 29–40.

Arrigo, Elisa. 2016. "Deriving Competitive Intelligence from Social Media." *International Journal of Online Marketing* 6 (2): 49–61.

Badr, A., E. Madden, and S. Wright. 2006. "The Contribution of CI to the Strategic Decision Making Process: Empirical Study of the European Pharmaceutical Industry." *Journal of Competitive Intelligence and Management* 3 (4): 15–35.

Badr, A., S. Wright, and D. Pickton. 2004. *Competitive Intelligence and the Formulation of Marketing Strategy*. Report by CIMITRI. Leicester, UK: Leicester Business School, De Montfort University.

Barnea, Avner. 2016. "Study on Competitive Intelligence in Israel." *Journal of Intelligence Studies in Business* 6 (2): 5–16.

Bulger, Nanette. 2016. "The Evolving Role of Intelligence: Migrating from Traditional Competitive Intelligence to Integrated Intelligence." *The International Journal of Intelligence, Security, and Public Affairs* 18 (1): 57–84.

Callegaro, Mario, Katja Lozar Manfreda, and Vasja Vehovar. 2015. *Web Survey Methodology*. 1st ed. London: Sage Publications.

Calof, Jonathan. 2014. "Want a Good Return for Your CI Dollar? Consider Event Intelligence." *Competitive Intelligence Magazine* 17 (4): 40–45.

Calof, Jonathan, Gregory Richards, and Jack Smith. 2015. "Foresight, Competitive Intelligence and Business Analytics — Tools for Making Industrial Programmes More Efficient." *Foresight-Russia* 9 (1): 68–81.

Calof, Jonathan, and Jack Smith. 2009. "The Integrative Domain of Foresight and Competitive Intelligence and its Impact on R&D Management." *R&D Management* 40 (1): 31–39.

Canongia, Claudia, Adelaide Antunes, and Maria de Nazaré Freitas Pereira. 2004. "Technological Foresight—the use of Biotechnology in the Development of new Drugs Against Breast Cancer." *Technovation* 24 (4): 299–309.

Competitive Intelligence Ning. 2016. "Competitive Intelligence." *Competitiveintelligence.Ning.Com*. http://competitiveintelligence.ning.com/.

Daheim, Cornelia, and Gereon Uerz. 2008. "Corporate Foresight in Europe: From Trend Based Logics to Open Foresight." *Technology Analysis & Strategic Management* 20 (3): 321–336.

Du Toit, A. 2013. "Comparative Study of Competitive Intelligence Practices between Two Retail Banks in Brazil and South Africa." *Journal of Intelligence Studies in Business* 2 (2): 30–39.

Du Toit, A. 2015. "Competitive Intelligence Research: An Investigation of Trends in the Literature." *Journal of Intelligence Studies in Business* 5 (2): 14–21.

Erickson, Scott G., and Helen N Rothberg. 2012. "Variation in Intellectual Capital Strategies Across Industries." In *4Th European Conference on Intellectual Capital*. Academic Publishing International.

Fehringer, Dale, Bonnie Hohhof, and Ted Johnson. 2006. *State of the Art Competitive Intelligence*. 1st ed. Washington: Competitive Intelligence Foundation.

Fleisher, Craig S., and Babette E. Bensoussan. 2003. *Strategic and Competitive Analysis*. 1st ed. Upper Saddle River, NJ: Prentice Hall.

Fleisher, Craig S., and Babette E. Bensoussan. 2012. *Business and Competitive Analysis*. 2nd ed. Upper Saddle River, NJ: Financial Times Press.

"FOR-LEARN.". 2017. *Forlearn.Jrc.Ec.Europa.Eu*. http://forlearn.jrc.ec.europa.eu/index.htm.

Fulton, B. R. 2016. "Organizations and Survey Research: Implementing Response Enhancing Strategies and Conducting Nonresponse Analyses." *Sociological Methods & Research*. http://journals.sagepub.com/doi/abs/10.1177/0049124115626169.

Global Intelligence Alliance. 2005. *Competitive Intelligence in Large Companies - Global Study*. GIA White Paper 4/2005. Finland: Global Intelligence Alliance.

Groves, Robert M., Floyd J. Fowler Jr., Mick P. Couper, James M Lepkowski, Eleanor Singer, and Roger Tourangeau. 2009. *Survey Methodology*. 2nd ed. Hoboken, NJ: John Whiley & Sons.

Hammoud, Mohamad S., and Douglas P. Nash. 2014. "What Corporations Do with Foresight?" *European Journal of Futures Research* 2 (1): 1–20.

Hitt, Michael A., R. Ireland, and R. Hoskisson. 2000. *Case Notes for Strategic Management: Competitiveness and Globalization. Annotated Textbook*. 4th ed. New york: Cengage Learning.

Jenster, P., and K. S. Soilen. 2013. "The Relationship between Strategic Planning and Company Performance–a Chinese Perspective." *Journal of Intelligence Studies in Business* 3 (1): 15–30.

Kahaner, Larry. 1996. *Competitive Intelligence*. 1st ed. New York: Simon and Schuster.

M-Brain. 2015. Global Market Intelligence Survey 2015. M-Brain White Paper, November 2015.

Neugarten, Michael L. 2006. "Foresight—are we Looking in the Right Direction?" *Futures* 38 (8): 894–907.

Olsmats, Carl, and Jari Kaivo-oja. 2014. "European Packaging Industry Foresight Study—identifying Global Drivers and Driven Packaging Industry Implications of the Global Megatrends." *European Journal of Futures Research* 2 (1): 1–10.

Peytchev, A. 2012. "Consequences of Survey Nonresponse." *The ANNALS of the American Academy of Political and Social Science* 645 (1): 88–111.

Plessis, Tanya du, and Mzoxolo Gulwa. 2016. "Developing A Competitive Intelligence Strategy Framework Supporting The Competitive Intelligence Needs of a Financial Institution's Decision Makers." *SA Journal of Information Management* 18 (2). Accessed 3 June 2017. http://www.sajim.co.za/index.php/SAJIM/article/view/726.

Qiu, Tianjiao. 2008. "Scanning For Competitive Intelligence: A Managerial Perspective." *European Journal of Marketing* 42 (7/8): 814–835.

Rogelberg, S. G., and J. M. Stanton. 2007. "Introduction: Understanding and Dealing with Organizational Survey Nonresponse." *Organizational Research Methods* 10 (2): 195–209.

Rohrbeck, René, Cinzia Battistella, and Eelko Huizingh. 2015. "Corporate Foresight: An Emerging Field with a Rich Tradition." *Technological Forecasting and Social Change* 101: 1–9.

Sewdass, Nisha. 2012. "Proposing a Competitive Intelligence (CI) Framework for Public Service Departments to Enhance Service Delivery." *SA Journal of Information Management* 14 (1). doi:10.4102/sajim.v14i1.491.

Vriens, Dirk, and Klaus Solberg Soilen. 2014. "Disruptive Intelligence." *Journal of Intelligence Studies in Business* 4 (3): 63–78.

Wright, S., A. Badr, and A. Weiss. 2004. "Competitive Intelligence Through UK Eyes." *Journal of Competitive Intelligence and Management* 2 (2): 68–87.

Wright, Sheila, and Jonathan L. Calof. 2006. "The Quest for Competitive, Business and Marketing Intelligence: A Country Comparison of Current Practices." *European Journal of Marketing* 40 (5/6): 453–465.

Fading memories of the future: the dissipation of strategic foresight among middle managers

David Sarpong and Daniël Hartman

ABSTRACT
Strategic foresight among middle managers is crucial, considering their responsibilities and authority vested in them in directing everyday organising. Emphasising practices as the locus of strategic foresight, we argue that imposed organising processes and bureaucratic routines may interact to dissipate the cultivation of strategic foresight among middle managers in their situated practice. Building on an explorative case analysis of a European sportswear retail company, our study highlights how top-down changes in organising processes may induce the dissipation of organisational 'foresightfulness'. We identify four dimensions emphasised by the new organising processes and their associated routines (rhetoric of legitimation, instrumental rationality, suppression of creative freedom, and the formulations of solutions in search of problems) which typify the observed patterns of foresight dissipation among middle managers. The study and its findings extend our understanding of contextual antecedents that could lead to the dissipation of strategic foresight among middle managers in organizing.

Introduction

Strategic foresight – i.e. the creative evaluation and (re-)configuration of sources of potentialities (and limits) into resources and productive outcomes – has long been promoted as an important element in the creation and capture of competitive advantage (Chia 2008; Heger and Rohrbeck 2012; Tsoukas and Shepherd 2004). The nurturing and leveraging of strategic foresight at both the individual and group level has been associated with adaptive learning (Bootz 2010; Rohrbeck and Schwarz 2013), the identification of opportunities for innovation (Sarpong and Maclean 2011; Könnölä, Brummer, and Salo 2007) and strategic renewal (Vecchiato 2012; Baden-Fuller and Volberda 1997). While the locus attribution of strategic foresight remains debatable, middle managers in particular, by virtue of their broad responsibilities and roles and the authority vested in them in directing everyday organising, are frequently promoted as strategic actors in the nurturing and cultivation of strategic foresight (See: Constanzo 2004; Constanzo and Tzoumpa 2010; Rouleau and Balogun 2011).

Drawing on the recent turn to practices in contemporary social theory, in reiterating the strategic importance of the actions of middle level managers in enabling strategic foresight, scholars have emphasised organising practices as the site of strategic foresight (Sarpong, Maclean, and Davies 2013; Sarpong and Maclean 2014; Cunha, Palma, and da Costa 2006; Waehrens and Riis 2010). They go further to explain how micro-interactions and the (re)production of often taken-for-granted organisational routines and practices (Feldman 2003; Pentland and Reuter 1994) could help us grasp not just the logic of strategic foresight, but also novel ways of theorising strategic foresight as a distributed capability (Cunha, Palma, and and Costa 2006; Sarpong and Maclean 2011; McKelvey and Boisot 2009). In short, organising practices do matter for the cultivation of strategic foresight. Despite the progress made by the existing literature on the contribution of middle managers to foresight, Appiah and Sarpong (2015) observed that there is limited research exploring how organisational context and antecedents may influence strategic foresight. In particular, there is sparse literature examining the extent to which new organising processes and routines in organisations interact to cultivate (or dissipate) strategic foresight among middle managers in practice. By 'organising processes' we refer to the formal and informal canonical rules and structures that prescribe, coordinate and govern situated practices and the 'acceptable way' work is done (Sarpong, Maclean, and Davis 2013). Moving forward, we contribute to this body of literature by taking a step towards a theory of the dissipation of strategic foresight in contemporary organising. To do this, we studied how the introduction of new organising processes and routines (what we call 'post acquisition system re-alignment') aimed at streamlining the operations in and improving the competitiveness of a newly acquired business, might inadvertently lead to the dissipation of strategic foresight in practice.

Empirically, we focus on how post-acquisition, top management changes to the organising processes of one European sports apparel retail company precipitated the decline of strategic foresight among the firm's middle managers. We delineate our findings into the categories of rhetoric of legitimation, separation of knowledge forms, suppression of creative freedom and the formulation of solutions in search of problems, as these four dimensions of the new organising processes gave rise to the observed patterns of foresight dissipation among middle managers. Our study offers two contributions: firstly, our study provides insight into how the work of middle managers may contribute to organisational 'foresightfulness'; secondly, we introduce and describe four dimensions of organising processes and routines which explain how strategic foresight could be precipitated in organising.

The paper is organised as follows: in the next section, we provide an overview of the concept of strategic foresight and the role of middle managers in contributing to organisational foresightfulness. Following this, we explain our research methodology and provide an overview of our case study organisation, data gathering and analysis. After this, we present the case study evidence showing how the introduction of new pervasive organising processes dissipated the potential of middle managers to enact 'foresightful' actions in their situated practice. We then conclude with a discussion of our findings and their relevance to theory and practice.

Strategic foresight among middle managers

Broadly conceived as a human capacity to identify, explore and exploit opportunities glossed over by others, strategic foresight has emerged as a skill required for successful organising. As an organising competence, strategic foresight helps to prepare the organisation for the future by broadening the boundaries of perception in strategy formulation, evaluation and implementation (Cunha, Da Cunha, and Clegg 2008; Rohrbeck 2012). Acknowledging the relations and linkages between strategy and the future, together with the relevance of foresight for value creation and capture, strategic foresight is frequently presented as a managerial competence (Mackay and Mckiernan 2004; Major, Asch, and Cordey-Hayes 2001), with a positive correlation to firm performance (Amsteus 2012). From this perspective, the foresight literature tends to focus on the 'legendary' founder (Harris and Ogbonna 1999), the 'great' CEO (Bezold 2010; McKelvey and Boisot 2009) and organisational elites or top management teams (MacKay 2009; Constanzo 2004), in accounting for organisational foresightfulness.

Meanwhile, a recent scholarly turn to the coal-face of organising has directed attention to middle managers as a strategic group of doers whose actions, activities and routines could have profound influence on organisational foresightfulness and entrepreneurial initiatives (Tsoukas and Shepherd 2004; Costanzo and Tzoumpa 2010). Zooming in on the micro-practices and routines which form an integral part of the day-to-day work of organisational members, this stream of work does not only draw attention to the strategic and symbolic contribution of middle managers in enabling foresight; it also delineates how strategic foresight could be harnessed and nurtured in contemporary organising (e.g. Sarpong, Bi, and Amankwah-Amoah 2016; Darkow 2015). Specifically, they argue that middle managers' integrative responsibilities for ordering, directing, learning and organising (Kerzner 2013; Sarpong, Maclean, and Alexander 2013) play an important role in developing their organisation's capacity to cope with unforeseen external factors (Schwarz 2009) and in facilitating learning within teams and departments (Constanzo and Tzoumpa 2010). Balogun and Johnson (2005), in arguing that foresight does not apply solely to senior management, suggest that strategic foresight among middle management is much more crucial because of their knowledge and perception of both the internal and external environments in which their organisations operate (Darkow 2015; Constanzo and Tzoumpa 2010). From this perspective, successful corporate foresight exercises such as scenario planning, value chain analysis and business war-gaming all rely on the contribution of middle management to the interpretation of external views about the future and how they could be converted into internal re-usable knowledge (Major, Asch, and Cordey-Hayes 2001).

In our view, the successful integration of strategic foresight in organising is determined by the active interaction between what Appiah and Sarpong (2015) referred to as 'ostensive' and 'performative' aspects of routines within a focal organisation. It involves the strategic integration of disparate and often competing blocs of knowledge which are products of organisational activities and routines, organised into representations of potential futures (Feldman 2003; Durand 2008). Yet we know that organising routines are themselves shaped by organising processes (Felin et al. 2012; Weick, Sutcliffe, and Obstfeld 2005). In this regard, we observe that organising processes – the formal and informal canonical rules and structures that prescribe, coordinate and govern the situated practices of middle managers – are a useful antecedent to the enactment of strategic foresight in organising (Sarpong and Maclean 2016). In this paper, we contribute to a better understanding of how organising processes, whether seen from a change- or stability- inducing perspective, could lead to the dissipation of strategic foresight among middle managers. This, we argue, is likely to be the case when changes in organising processes are not well managed to accommodate feedback from middle managers and the organisation's external environment. Thus the main research question driving our empirical inquiry is therefore as follows: 'How can organising processes dissipate strategic foresight among middle managers?' Our empirical research context is a Dutch retail company that made sea changes to its organising processes after going through ownership changes. In this context, the company offers an interesting empirical research case study to examine the influence of organising processes on managerial foresight. We accorded the organisation in question a pseudonym (Mabodia Inc.) To preserve its anonymity. In the next section, we provide an overview of the case study.

The case study- mabodia inc

Mabodia Inc. is a sports apparel retailer based in the Benelux region. Mabodia Inc. was formed by a young entrepreneur and sport enthusiast some 30 years ago. It expanded by opening new stores, extending its collection and focussing on quality garments, personal advice, expert knowledge and service. By 2004, Mabodia Inc. had 27 stores in the Benelux region, a time when its growth and sales figures were attracting the attention of many investors. It also received great acclaim from the Dutch and Belgian press for its family-friendly flexible working practices, emphasis on creativity and managerial autonomy in making localised decisions. In that very year, an ex-employee, still enthusiastic about the company, swiftly initiated the acquisition of Mabodia Inc. by Kingtana, a Benelux sports investment company. The new parent firm decided to maintain the management of Mabodia Inc. and therefore made few changes to the business's organising processes.

In 2007, the company was acquired by a British private equity firm and became part of a group of sports companies. The new owners initiated some pervasive structural changes to the way work and business as a whole were organised at Mabodia Inc., to bring it into line and harmony with how 'things were done' in the other businesses within the group. The external image was revamped (new logos, new mission statement, new shop lay-outs). Among the major internal changes was the introduction of a monthly cost and turnover target for every store. The job and role of store managers was re-prescribed in detail and what store managers were expected to do in an 'emergency' was formalised and codified. In an effort to improve performance and efficiency, the planning of work rotas was transferred to the head office. No one was allowed to work extra hours. Employees were no longer to be paid for a forfeited break; hence store managers needed to make sure their staff took their breaks even when the store was busy. The corporate style of the J-60 parent group is one that emphasises planning and tight budget controls. Table 1 is a summary of the biographical case history of Mabodia Inc.

The changes, according to the company were completed three years after the acquisition and led to a major improvement in performance. Nevertheless, Mabodia Inc., which used to be an industry pace-setter, is currently posting just above industry growth rates. Management tend to blame the economically difficult time in which the firm is operating. However, according to store managers, while the new organising processes may be delivering profits for the organisation in the short term, the store managers no longer feel empowered to organise work. The shift from bottoms-up, empowered management closely attuned to the local market to a more top-down, numbers-driven management approach has been a bone of contention. At the time of data collection, we were told that staff turnover was an unprecedented level, and that many managers felt disenchanted about the new regime which tended to understand organisational performance only in terms of the balance sheet.

Research methodology

In addition to publicly available materials on Mabodia Inc., we searched the company's website and intranet and downloaded sets of texts and documents. These included organisational and reporting structures, strategy plans and newsletters. This gave us an unprecedented insight into the old and 'new ways' of organising business at Mabodia Inc. Our chosen level of analysis was the middle managers (store managers) in Mabodia Inc., who are relied upon by the organisation to use their skills, expertise and experiential knowledge to mobilise distributed organisational resources in order to get work done. From this pre-given ontological status, we conceptualise a typical Mabodia Inc. store manager as a 'manager who is forced to superficiality by fragmented, discontinuous events occurring throughout the organisation; the manager who must be a master politician, master negotiator, expert psychologist and counsellor, expert communicator, and much more' (Introna 1997, 22).

Due to practical considerations including availability and geographical proximity, we invited twenty managers, who had been in post long before the acquisitions, to take part in the study; fifteen of the twenty agreed (about 50% of Mabodia Inc. store managers). We then collected 10 narrative interviews from the store managers. The managers were interviewed over a period of four months. The twelve managers had over one hundred years' combined experience working at Mabodia Inc. and an average of seven years each as store managers. The interviews were open-ended, starting with broad questions about their everyday work and personal stories within

Table 1. Biographical case history of Mabodia Inc.

Founded by a sports enthusiast	Acquired by Kingtana Sports investment company	Acquired by a foreign equity firm – the J60-Group
• Privately owned firm • Simple management structures • Store managers played central role in decisions making	• Top management team maintained • Firm saw few changes in the way work was organised • Store managers' role in decision making was reduced	• Introduction of new management • New organising processes aimed at formalising and professionalising working practices • Store managers' role in decision making reduced to the barest minimum
Period 1: Late 1970s–2004	Period 2: 2004–2007	Period 3 2007–Present

Mabodia Inc. We then drilled further down to their perceptions of the acquisition transitions, how they have impacted on their situated practice and how work is organised in their individual stores. Each interview lasted approximately two hours. They were digitally recorded and transcribed within 24 h. In total, we generated 200 pages of interview transcripts.

The full data analysis followed four stages. First, we meticulously sifted through the interview data, then collated what we thought were salient narratives and cross-checked them with other vital objective information we had gleaned from the documentary sources, in order to overcome possible biases in what we heard in the field. Second, the disparate data were triangulated into a whole (Jick 1979), re-read with the aim of identifying recurrent comparative phrases which were noted and highlighted. Third, in comparison with the extant literature on strategic foresight, we analysed and interpreted the data until common themes started to emerge and became saturated (Corbin and Strauss 2008). In doing this we focussed on how the new organising practices introduced by top management significantly altered everyday organising practices and routines in the various stores. We used this to generate analytical categories which converged on four distinct but interrelated top level codes: rhetoric of legitimation, instrumental rationality, suppression of creative freedom, and grand solutions in search of problems. Finally, we applied the analytical categories to the entire data set, after which the emerging patterns were then used to develop greater insight and form descriptive explanations of the influence of the new organising practices on the dissipation of foresight potential among the store managers. Table 2 is an overview of our data analysis.

In the next section, we present how the vicissitudes of the current organising processes and the consequent assemblages of processes could lead to the dissipation of strategic foresight among the middle managers who are supposed to be the protagonists of the future.

When the organising processes unravels

According to accounts from inside the organisation, the formalisation of the organising processes had the good intention of professionalising the organisation and its workforce. However, the zeal

Table 2. Overview of data analysis.

First-order codes	*Theoretical categories*	*Aggregate theoretical dimension*
▪ Strict benchmarking of new organising processes ▪ Rules discounting old organising processes ▪ Emphasis on bureaucratic routines	Rhetoric of legitimation	*Dissipation of strategic foresight in organizing*
▪ Over-emphasises on scientific and technical knowledge ▪ Overt discounts of experiential knowledge	Instrumental rationality and the separation of knowledge forms	
▪ Codification of actions in response to 'breakdowns' in practice ▪ Broadly defined sanctions against organising deviance ▪ Resource rationing	Suppression of creative freedom	
▪ Managerial guidelines on how to organise the work of junior employees ▪ Preconceived and set-in-stone plans from the head office ▪ Forced synergies between activities	Grand solutions in search of problems	

of management in changing the organising regimes means they ignored the co-dependency of the new formalities and the long existing informal routines that had served and sustained the organisation for decades (Marlow, Taylor, and Thompson 2010). This, we argue, led to overt contestation of the new organising processes. In order to get middle managers to accept the new organising processes, the organisation then went on to launch a series of campaigns to persuade these managers to get on board. The outcome of the campaign manifested in four closely related forms of activities that gave rise to the dissipation of strategic foresight among the middle managers. We present these organising processes under four salient rubrics: rhetoric of legitimation, instrumental rationality, suppression of creative freedom and the formulation of solutions in search of problems.

The rhetoric of legitimating the new organising processes

In their efforts to streamline the operations of their newly acquired business unit, the new management introduced a raft of changes to the organising processes and routines in order to bring the company in line with their existing business units. A participant described one particular feature of the new organising processes:

> Currently the focus is on up-selling of maintenance products; for instance with shoes. It only counts as up-selling when the maintenance products are on the same receipt, but what we see is that people think about it and buy the maintenance products later. They will! But head office does not look further than the receipts … [MM8].

Targets were set for every store, and new reporting procedures were imposed. In the eyes of the middle managers, the new organising processes which were solely driven by 'targets' were de-skilling in nature, and reduced their autonomy in making decisions about how work was organised in their stores.

> … these targets are used as benchmarks; for the head office, these targets were important and had to be met. Since we are evaluated on this, we continuously focus on these targets and manage our personnel in not so pleasant ways to hit the targets. [MM4]

The middle managers who were not very pleased with the new changes raised concerns with the new head office and some went even further, filing a series of complaints about how their accumulated experiences at the individual level were being truncated by the over-emphasis on efficiency. Rather than confronting the concerns raised, management chose to promote the new organising processes as the ultimate means of sustaining the business. Backed with statistics that 'problematise' the past organising processes, employees were explicitly made to understand the old ways of organising work in the stores were not only unproductive but also arcane and retrogressive. Few resigned and those who proved deviant were dismissed. As observed by one participant:

> All they were interested in was to run the company as efficiently as they can and to get the most out of it in terms of profit. They will do it at the expense of staff … and you cannot challenge [MM2].

The rhetoric used by management to legitimise the new organising processes reinforced the dichotomy between management and the middle managers about their visions of the future. Feldman (2003) argues that organisational actors in practice frequently use their understanding of how their organisations operate as a benchmark to guide their performances within their situated routines and this suggests that agency shapes the performance of organising routines, which in turn drives the enactment of foresightful actions in practice.

> Everyone felt appreciated for what they used to do; the current message from head office however, is that 'it must go better in the future', which can be harsh after you tried your best and have made the improvements they required [MM8].

In summary, we assert that through the imposition of the new organising regime aimed at efficiency gains, management tends to undermine the store managers' accumulated experiential knowledge.

While the gaps between the new and old regimes could potentially lead to the identification of opportunities for innovation, for example, we argue that the zeal with which the old processes are discredited by top management deters store managers from engaging in activities which could contribute to the creative search for opportunities and subsequent exploration.

Instrumental rationality and the separation knowledge

Our data evidence also suggests that the new organising processes imposed by management placed a lot of emphasis on rationality at the expense of the intuition which drives strategic foresight in organising. This involved the active separation of the technical and embodied knowledge drawn upon by middle managers in their everyday organising and the interpretation of the usefulness of newly developed knowledge in practice. The evolving conceptual wedge between these bodies of knowledge became filled with audits and performance benchmarks. As a result, the organising process disempowers managers to spontaneously adapt their staffing arrangements into meaningful social order, especially when there is a temporary breakdown. The observation by one manager in particular reflects such visceral experiences:

> Head office even suggests how to plan my staffing for the next month, without knowing the exact circumstances of my store. Their view is solely based on figures of the last year and how to improve those [MM8].

In such a situation, while this manager has an agentic responsibility of prescribing work for junior employees, the over-prescriptive organising processes to which they must adhere end up overriding their improvisation potential to deal with problems within the contingency of the moment. In a related development, our store managers reported that they plan the future work in their stores with a maximum time-horizon of one year. This horizon, we inferred, is instigated by head office, which releases a 'retail calendar' including a general prognosis and plan for the next year. One manager explains:

> … it is evolving; is becoming more figure-based, I mean so far we are looking at turnover and staffing costs as the two main variables, but it is becoming more figure orientated. [MM1].

Surprisingly, the store managers tend view this calendar as 'vague' and 'incomplete'. Reinforcing the gulf between management, and the store managers, they all conceded they hardly have the chance to analyse the future beyond three months. Clearly, while they are well aware that unforeseen developments, such as the weather, the market itself and personnel availability could alter the future and even the present, they were constrained by the strict formal planning regime within which they were expected to operate. One manager lamented:

> Nowadays the (middle) managers are more like pawns, instead of the player that controls the game. [MM9]

Apart from the strict planning regime, our middle managers also felt the new organising regime over-prioritises bureaucratic routines at the expense of their well-being.

> Before the change of ownership the CEO could walk into your store, ask you how you doing, including your pet. When the new CEO walks in, there is no greeting and he immediately starts talking about things he does not like in your store and wants to be changed [MM4].

While the new CEO reserves the right not to inquire about employees' life outside work, most of the store managers reported that they used to see the business as a family and yearn for the good old days when you know the boss cared about not just about the bottom line, but also the well-being of employees. In this regard, we co-observed narratives about what it used to be like to work for the organisation and what could have been accomplished if the culture had been protected. These narratives, we observed, were being used by managers to resist the new organising demands. The offshoot is a feeling of threatened identity and the urge not to 'respond to circumstances so that the organisation could get around in the world' (Tsoukas and Shepherd 2004, 10).

Suppressing creative freedom

Creativity lies at the heart of strategic foresight. It drives visioning, proactive learning established around continuous probing, interpretation of equivocal cues and the integration of disparate ideas to develop images of the future (Constanzo 2004). Our case study evidence suggests that the new organising process which was very much set in stone, gave store managers little space to reflect on their work or to challenge their own taken-for-granted assumptions about how work should be organised. They were expected to follow laid down procedures in dealing with 'breakdowns' and their input into decisions about how their stores should be run was reduced to the barest minimum. Unfortunately, the intended strategy led to unintended outcomes in terms of creativity and the overall sense-making potential of managers (Balogun and Johnson 2005). One store manager said she felt that she was now more of an executor than a manager. She went on to explain:

> Currently, my role in the store is completely different; with the previous owner I was empowered to make decisions, and of course was held responsible for the outcomes. Now I just have to do what head office tells me, and they still hold me responsible for the outcomes; even when I initially did not liked the idea ... [MM10]

Store managers are seen here to have far less agency and control, not just about strategic decisions but about the everyday running of their shops. They are expected to get clearance from the head office before making very basic decisions that may require immediate attention within the contingency of organising. One manager described what happened when some devices in his store broke down:

> I sent them (the head office) a couple of emails about a broken door, but they did not respond. They only got back to me when I sent them the quotation for the repairs. They reacted by saying that I was not allowed to do things without their consent or approval [MM3].

The above quote suggests there is little room for store managers to use their intuition or past experience to resolve organising bottlenecks (Akinci and Sadler-Smith 2012). Note that management did not only stop this store manager from going ahead to get the door repaired, they went further, reminding him that he cannot make routine decisions until he has been advised when and how to resolve a problem. In this case, for example, we argue that the action of the employer undermines managers' initiative and their willingness to embrace their responsibilities as a challenge to be mastered (Sarpong, Maclean, and Davies 2013) and typically results in managers acting too late in dealing with problems arising within the contingency of organising. The potential consequence is that those who have invested their efforts and reputations in the organisation begin to frame everyday improvisation or a call to duty beyond their immediate responsibilities as a threat to be avoided. This was the case when another manager observed that:

> I used my experience in planning to anticipate what I expected to happen in the coming period. I did so automatically, but with the focus on meeting targets, I noticed I have little time thinking about the future [MM10].

Here we suggest that the prevalence of the unravelling organising processes has damaging effects not just on the identification and evaluation of opportunities for learning and innovation: it also stifles strategic thinking and the creative potential of store managers.

Grand solutions in search of problems

The J-60 group had made it clear to all employees they were keen to improve operational efficiency when they took over Mabodia Inc. As observed by most of our store managers, rather than looking for challenges with the old organising processes and then looking for ways to improve them, the new owners were more interested in bringing Mabodia Inc.'s operations into line and harmony with the other businesses within the J-60 Group. This involved making changes to processes with no regard to context. Over-reliance on rational analysis in order to design a fit between Mabodia Inc. and other businesses led to forced synergies between many organising activities and routines. As recounted by one store manager, some of the changes were simply not worth the bother of implementing:

> Although the general planning of the head office offers some guidance, it does happen that most of the information they provide are wrong or incomplete; school holidays are wrongly planned or completely forgotten about ... and since holidays are peak-moments for us, it is of utmost importance to know when to expect them, and prepare accordingly. [MM7]

Store managers had no input into these plans, which were normally delivered to store managers as spreadsheets. Per their new operational guidelines, they were not expected to engage with these spreadsheets, all they needed to do was to fill in the blanks, which the head office can then use for further planning. The major problem associated with this 'spreadsheet management' is the inherent lack of flexibility built into operations:

> There is a lot of planning involved, but there is also a lot of reaction involved if you want the business to do well, and my staffs do understand. So for example, if it freezes for two weeks in January and it snows, you basically have to throw everything out of the window and get everybody to work sixty hours in the week to make sure you take advantage of the situation. Unfortunately, Head office just can't understand that if you want to achieve results you have to be flexible [MM2].

As narrated by this manager, their work (which was now planned in detail by head office) was extremely inflexible. Breaks were now compulsory and had to be taken at a pre-specified time, even if the shop was busy. In addition, no one was expected to work extra hours, a formal measure which, he argued, was unnecessarily targeted by top management to reduce staff costs. Each store was also given a monthly performance target with associated managerial guidelines on how to organise work in that particular month. Nine of our store managers found these quantitative targets problematic because what they seek to measure can, in essence, sometimes be unachievable, unrealistic or even un-measurable. One manager explains how some of these targets cannot be met due to circumstances the head office is not aware of:

> You see, sometimes the store layout has to be changed; resulting in the closure or limited accessibility of some departments, or when the winter sports department is transformed to a camping department. That could take up to two weeks, resulting in a decline of sales in that period. This is necessary for the sales of that whole season, but you are still asked to explain why you didn't meet targets during the transformation period [MM5].

As recounted by the manager, the targets might not be met in the short term, but the overall turnover of the store, as a result of the ad hoc transformation, might be increased in the long-run. However, the micro-management of the organising processes means store managers get fixated on meeting set-in-stone quantitative targets. In this regard, store managers are seldom able to balance their short-term performance imperatives with the long-term 'foresightful' needs of the organisation. In summary, they refuse to 'think the unthinkable' and expend little time in reflecting on their practice.

Discussion and conclusions

Strategic foresight is imperative for successful organising. In this paper, we examined how organising processes could potentially dissipate the nurturing of strategic foresight among middle managers. We did this by studying how a new organising regime imposed on a successful sports apparel retail firm led to the dissipation of foresight among its middle managers. By drawing on a performative practice approach in our study of strategic foresight among middle managers, we bring in a wider array of influences to the theory and practice of strategic foresight, showing how organising processes, as a causal factor lying outside language, may influence talk, actions and, in turn, the enactment of strategic foresight by middle managers.

Our case evidence suggests that the acquisition of Mabodia Inc. meant that the company was now a strategic business unit of a multi-business company, J-60. The parent company, in an effort to justify to itself that its influence would lead to better performance in Mabodia Inc. than it would otherwise achieve as an independent, stand-alone entity, introduced new organising processes explicitly aimed at formalising and professionalising working practices. These changes, we observe, led to worthless, bureaucratic routines and a sharper concentration of everyday decision making in the hands of top management. The new organising regime therefore resulted in radical changes to the organising architecture, which in turn transformed how work was organised and the responsibilities, activities and routines of store managers. Practices that were once normal disappeared overnight. We found that such a new organising regime disempowered store managers in making decisions within the contingency of organising. In response, store managers became 'mindless-in-practice', thus, their capacity to sensitivity and awareness (Chia 2008) was dulled to the extent that they not only expended little effort in identifying opportunities and potentialities to improve their work, but also unreflectively waited for instructions from above to manage basic tasks.

Our findings suggest that when faced with the need for empowerment, responsiveness to local demands, and value capture for shareholders, the management chose profits over future orientation. This shows that while organisations are frequently preoccupied with making changes to their organising practices, paradoxically, they are less able to manage the influence of such changes on the future orientation of the organisation as a whole. This raises an important question that lies at the core of foresight: what is the value of (long-term, qualitative) foresight in the context of (short-term quantitative) financial performance? Our research therefore has some implications for practice. Cunha, Palma, and and Costa (2006) argue that everyday micro-interactions, foresightful micro-behaviour and the actions of middle managers are important in defining the future of an organisation. In this regard, when upper management decide to convert organising processes into quantitative calculable and saleable units, then, of course, middle managers will feel trapped in an 'organising cul-de-sac' where their perceptual abilities and improvisation skills, for example, are evaluated and measured by box ticking. Caught in such an organising milieu, both the enactment of everyday foresightful actions and the cultivation of strategic foresight are likely to be dissipated among middle managers. While the findings of our study are illuminating, there is a question as to whether the organising patterns and dynamics reported in our study are a simple manifestation of normative acquisition outcomes. In this regard, the insights from our study could serve as a first step towards stimulating additional research into the influence of organising practices on organisational 'foresightfulness'. In accounting for foresight in the form of strategising, our reliance on textual data gave us little room to show, if indeed it were possible, how organising processes could be observed and theorised outside language. Future research may employ ethno-methodologies such as participant observation, work shadowing, video-ethnography or action research to potentially capture data on organising processes which could possibly impede middle managers' enactment of foresightful actions in their situated practices.

For the present, we are not so bold as to argue that the new organising processes are having a detrimental impact on the competitiveness of the organisation, especially when the firm is currently reporting above industry profits. The fact that the company is now realising above-industry financial results suggests positive short-term improvements; however, our interviews with store managers suggest that the financial focus is too intense, and already is having negative effects on employee empowerment and the stores responsiveness to local conditions. In the words of Alfred North Whitehead:

> The motive of success is not enough. It produces a short-sighted world which destroys the sources of its own prosperity … A great society is a society in which its men of business think greatly about its function. Low thoughts mean low behaviour, and after a brief orgy of exploitation, low behaviour means a descending standard of life (Whitehead [1932] 1933, 120).

Whiteheads observation raises some pertinent questions related to what counts as best practice to organizing in fast changing environments, the locus of knowledge, and power in organisations: is the new leadership too extreme in implementing the new approach? Are they 'bad' managers – or, in fact, might they be prescient, recognizing important shifts in the industry in that the 'old ways' of doing things are no longer relevant? Again, this judgment is beyond the scope of this paper.

Nevertheless, we are of the view that the continuous dissipation of strategic foresight among middle managers may have an adverse impact on the organisation's ability to leverage the potential of its human resources, thereby inhibiting the organisation from surpassing established boundaries or seising the opportunities overlooked by competitors.

Disclosure statement

No potential conflict of interest was reported by the authors.

References

Akinci, C., and E. Sadler-Smith. 2012. "Intuition in Management Research: A Historical Review." *International Journal of Management Reviews* 14 (1): 104–122.

Amsteus, M. 2012. "A Valid Matter for Managerial Foresight." *Journal of Futures Studies* 17 (2): 59–74.

Appiah, G., and D. Sarpong. 2015. "On the Influence of Organisational Routines on Strategic Foresight." *Foresight (los Angeles, California)* 17 (5): 512–527.

Baden-Fuller, C., and H. W. Volberda. 1997. "Strategic Renewal: How Large Complex Organisations Prepare for the Future." *International Studies of Management & Organisation* 27 (2): 95–120.

Balogun, J., and G. Johnson. 2005. "From Intended Strategies to Unintended Outcomes: The Impact of Change Recipient Sensemaking." *Organization Studies* 26 (11): 1573–1601.

Bezold, C. 2010. "Lessons From Using Scenarios for Strategic Foresight." *Technological Forecasting and Social Change* 77 (9): 1513–1518.

Bootz, J. P. 2010. "Strategic Foresight and Organisational Learning: A Survey and Critical Analysis." *Technological Forecasting and Social Change* 77 (9): 1588–1594.

Chia, R. 2008. "Enhancing Entrepreneurial Learning Through Peripheral Vision." In *Entrepreneurial Learning: Conceptual Frameworks and Applications*, edited by R. T. Harrison and C. M. Leitch, 27–43. London: Routledge.

Constanzo, L. 2004. "Strategic Foresight in High-Speed Environment." *Futures* 36 (2): 219–235.

Constanzo, L. A., and V. Tzoumpa. 2010. "The Role of Middle-Managers in Enabling Foresight". In *Handbook of Research on Strategy and Foresight*, edited by L. A. Constanzo and R. B. Mackay, 277–296. Cheltenham: Edward Elgar.

Corbin, J., and A. Strauss. 2008. *Basics of Qualitative Research*. 3rd ed. Thousand Oaks: Sage.

Cunha, M. P., J. V. Da Cunha, and S. R. Clegg. 2008. "Improvisational Bricolage: A Practice-Based Approach to Strategy and Foresight." In *Handbook of Research on Strategy and Foresight*, edited by L. A. Costanzo and R. B. Mackay, 182–199. Cheltenham: Edward Elgar.

Cunha, M. P., P. Palma, and N. G. da Costa. 2006. "Fear of Foresight: Knowledge and Ignorance in Organisational Foresight." *Futures* 38 (8): 942–955.

Darkow, I. L. 2015. "The Involvement of Middle Management in Strategy Development— Development and Implementation of a Foresight-Based Approach." *Technological Forecasting and Social Change* 101: 10–24.

Durand, T. 2008. "Scenarios as Knowledge Transformed Into Strategic 'Re-presentations': The Use of Foresight Studies to Help Shape and Implement Strategy." *Management & Avenir* 17: 279–297.

Feldman, M. S. 2003. "A Performative Perspective on Stability and Change in Organisational Routines." *Industrial and Corporate Change* 12 (4): 727–752.

Felin, T., N. J. Foss, K. H. Heimeriks, and T. L. Madsen. 2012. "Microfoundations of Routines and Capabilities: Individuals, Processes, and Structure." *Journal of Management Studies* 49 (8): 1351–1374.

Harris, L. C., and E. Ogbonna. 1999. "The Strategic Legacy of Company Founders." *Long Range Planning* 32 (3): 333–343.

Heger, T., and R. Rohrbeck. 2012. "Strategic Foresight for Collaborative Exploration of New Business Fields." *Technological Forecasting and Social Change* 79 (5): 819–831.

Introna, L. D. 1997. *Management, Information and Power: A Narrative of the Involved Manager*. Basingstoke: Macmillan Press.

Jick, T. D. 1979. "Mixing Qualitative and Quantitative Methods: Triangulation in Action." *Administrative Science Quarterly* 24 (4): 602–611.

Kerzner, H. 2013. *Project Management: A Systems Approach to Planning, Scheduling, and Controlling*. John Wiley & Sons.

Könnölä, T., V. Brummer, and A. Salo. 2007. "Diversity in Foresight: insights from the Fostering of Innovation Ideas." *Technological Forecasting and Social Change* 74 (5): 608–626.

Major, E., D. Asch, and M. Cordey-Hayes. 2001. "Foresight as a Core Competence." *Futures* 33 (2): 91–107.

MacKay, B. 2009. "Strategic Foresight: Counterfactual and Prospective Sense Making in Enacted Environments." In *Handbook of Research on Strategy and Foresight*, edited by L. A. Costanzo and R. B. Mackay, 90–112. Cheltenham: Edward Elgar.

MacKay, R. B., and P. McKiernan. 2004. "Exploring Strategy Context with Foresight." *European Management Review* 1 (1): 69–77.

Major, E., D. Asch, and M. Cordey-Hayes. 2001. "Foresight as a Core Competence." *Futures* 33 (2): 91–107.

Marlow, S., S. Taylor, and A. Thompson. 2010. "Informality and Formality in Medium-Sized Companies: Contestation and Synchronization." *British Journal of Management* 21: 954–966.

McKelvey, B., and M. Boisot. 2009. "Redefining Strategic Foresight: 'Fast' and 'Far' Sight via Complexity Science." In *Handbook of Research on Strategy and Foresight*, edited by L. A. Costanzo and R. B. Mackay, 15–47. Cheltenham: Edward Elgar.

Pentland, B. T., and H. H. Reuter. 1994. "Organisational Routines as Grammar of Action." *Administrative Science Quarterly* 39: 484–510.

Rohrbeck, R. 2012. "Exploring Value Creation from Corporate-foresight Activities." *Futures* 44 (5): 440–452.

Rohrbeck, R., and J. O. Schwarz. 2013. "The Value Contribution of Strategic Foresight: Insights from an Empirical Study of Large European Companies." *Technological Forecasting and Social Change* 80 (8): 1593–1606.

Rouleau, L., and J. Balogun. 2011. "Middle Managers, Strategic Sensemaking, and Discursive Competence." *Journal of Management Studies*, 48 (5): 953–983.

Sarpong, D., J. Bi, and J. Amankwah-Amoah. 2016. "On the Nurturing of Strategic Foresight: The Ubuntu Perspective." *Futures* 75: 14–23.

Sarpong, D., and M. Maclean. 2011. "Scenario Thinking: A Practice-Based Approach for the Identification of Opportunities for Innovation." *Futures* 43 (10): 1154–1163.

Sarpong, D., and M. Maclean. 2014. "Unpacking Strategic Foresight: A Practice Approach." *Scandinavian Journal of Management* 30 (1): 16–26.

Sarpong, D., and M. Maclean. 2016. "Cultivating Strategic Foresight in Practice: A Relational Perspective." *Journal of Business Research* 69 (8): 2812–2820.

Sarpong, D., M. Maclean, and E. A. Alexander. 2013. "Organizing Strategic Foresight: A Contextual Practice of 'Way Finding'." *Futures* 53 (9): 33–41.

Sarpong, D., M. Maclean, and C. Davis. 2013. "Matter of Foresight: How Organising Practices Enable (or Impede) Organisational Foresightfulness." *European Management Journal* 31 (6): 613–625.

Schwarz, J. O. 2009. "Business Wargaming: Developing Foresight Within a Strategic Simulation." *Technology Analysis & Strategic Management* 21 (3): 291–305.

Tsoukas, H., and J. Shepherd. 2004. *Managing the Future: Foresight in the Knowledge Economy*. Oxford, UK: Blackwell Publishing.

Vecchiato, R. 2012. "Environmental Uncertainty, Foresight and Strategic Decision Making: An Integrated Study." *Technological Forecasting and Social Change* 79 (3): 436–447.

Waehrens, B. V., and J. O. Riis. 2010. "Failures to Enact the Future: A Social Practice Perspective." *Futures* 42 (4): 328–336.

Weick, K. E., K. M. Sutcliffe, and D. Obstfeld. 2005. "Organizing and the Process of Sensemaking." *Organization Science*, 16 (4): 409–421.

Whitehead, A. N. (1932) 1933. *Adventures of Ideas*. Harmondsworth: Penguin.

Participation in inter-organisational collaborative open foresight A matter of culture

Melanie Wiener, Regina Gattringer and Franz Strehl

ABSTRACT

Foresight at the fuzzy front end of the innovation process has become a widespread managerial practice to increase innovation capacity. Nevertheless, conducting foresight by focusing exclusively on internal know-how and capabilities might inhibit companies from maximising the full potential of this managerial practice. Therefore, opening up companies' boundaries to look into the long term future jointly with other organisations and to share knowledge on possible future developments, is proposed as a promising approach against these limitations. Before committing to inter-organisational collaborations like collaborative open foresight, it is important to analyse whether the company possesses the necessary prerequisites. Findings from adjacent areas indicate that organisational culture and its influence on a company's openness to engage in open approaches is an important precondition. So far, no clear evidence of a culture fostering openness for open foresight has yet been presented. Based on two in-depth case studies applying the Competing Values Framework for (organisational) culture analysis we identify cultural values that foster or inhibit openness: (1) A corporate culture characterised by a clan culture and adhocracy culture is fostering openness, whereas (2) a market and hierarchy culture inhibits openness for open foresight participation.

Introduction

In a business environment which is characterised by continuous, often disruptive change (Vecchiato and Roveda 2010; Weber et al. 2012), companies face increasing pressure to identify future opportunities that enable them to create new products, services and markets and retain their competitive capability in the future (Boer and Gertsen 2003; Fink, Lang, and Harms 2013; Huang et al. 2012; Sarpong, Maclean, and Davies 2013). Therefore, a growing number of companies see foresight as an important tool for preparing for the future (Daheim and Uerz 2008; Sarpong, Maclean, and Davies 2013).

Several authors argue that especially in the fuzzy front end of the innovation process, foresight plays an important role (Rohrbeck 2014; Van der Duin and Den Hartigh 2009). For example Rohrbeck (2014, 72) claims that '*By applying foresight methods, firms are able to channel more future insights into their front end of innovation and thus increase the likelihood of discovering interesting opportunities.*' Furthermore, Van der Duin and Den Hartigh (2009) state that the future can be a source of inspiration and should be taken into account in every innovation process.

Nevertheless, conducting foresight by focusing exclusively on internal know-how and capabilities might inhibit companies from finding significant new ideas (Hansen, Bullinger, and Reichwald 2011;

Keinz and Prügl 2010). Therefore, similar to the open innovation paradigm (Chesbrough 2003), foresight developed to open foresight – *'a process for analysing complex issues in an open and collaborative way'* (Miemis, Smart, and Brigis 2012, 92).

It has to be mentioned that the idea of integrating external sources is not new. The importance of network approaches (platforms for exchange e.g. World Futures Studies Federation, World Future Society, Global Business Network, Millennium Project) and the consideration of external resources were recognised very early in the history of foresight (Ehls et al. 2016; Ramos, Mansfield, and Priday 2012). However, it is argued that the open foresight approach better integrates the idea of connectivity and the use of various external sources (Ehls et al. 2016). In addition, diverse challenges – e.g. increasing environmental complexity and fast technological change – and the requirement to consider a large variety of perspectives and data sources force companies to reshape organisational practices and to collaborate with other companies (Paliokaitė, Pačėsa, and Sarpong 2014; Rohrbeck, Battistella, and Huizingh 2015; Van der Duin, Heger, and Schlesinger 2014). This leads to an increasing interest in inter-organisational collaborations when conducting foresight in the past few years (Heger 2014). Rohrbeck, Battistella, and Huizingh (2015) even claim that the importance of collaborations in the field of foresight is continuing to grow.

In this study we focus on collaborative open foresight which means a joint analysis and discussion process of various organisations concerning future developments in specific search fields. This promising approach is supposed to foster collection of broader data, sharing of expertise and resources, inclusion of new perspectives from different backgrounds and 'out-of-the-box-thinking' (Burmeister and Schulz-Montag 2009; Heger and Boman 2015; Keinz and Prügl 2010; Van der Duin, Heger, and Schlesinger 2014). For example, Gattringer, Wiener, and Strehl (2017) show – based on an action research study – that collaborative foresight (conducted by five companies) was fruitful for the creation of new perspectives and ideas and fostered 'out-of-the-box-thinking'. Through collaborative open foresight, know-how about the future can be developed together and then used on a single company level (Burmeister and Schulz-Montag 2009; Keinz and Prügl 2010).

However, moving to these collaborations across companies' boundaries is not a trivial task (Hansen and Klewitz 2012; Remneland-Wikhamn and Wikhamn 2011) as culture between open and closed organisations differs (Herzog and Leker 2010; Van de Vrande et al. 2009) and requires a different way of thinking (Van der Meer 2007). Therefore, it is important to analyse whether the company possesses the necessary preconditions – especially sufficient openness.

Openness of corporate culture is not only said to be a prerequisite for open foresight (Heger and Boman 2015), but also for various forms of open innovation (Brettel and Cleven 2011; Docherty 2006; Enkel, Bell, and Hogenkamp 2011; Giannopouou, Yström, and Ollila 2011; Herzog and Leker 2010; Remneland-Wikhamn and Wikhamn 2011). Our understanding of openness is an 'outside-in' approach and is defined as *'a general willingness to access new knowledge, a positive attitude towards complementary influence, and, consequently, the willingness to exchange with others'* (Enkel 2010, 349). To favour this kind of innovative, open culture a special set of attitudes and values is necessary (Stock, Six, and Zacharias 2013; Van der Meer 2007). Nevertheless, this question of an *'appropriate culture required for the implementation of an open approach to innovation'* (Mortara et al. 2010, 370) is largely unaddressed in the literature (Brettel and Cleven 2011; Lahi and Elenurm 2015; Szymańska 2016).

Based on the call for further research in this regard from adjacent areas as well as from open foresight researchers (Ehls et al. 2016) we investigate the following research question: Which features of corporate culture foster or inhibit a company's openness for collaborative open foresight?

Drawing on insights from foresight and (open) innovation management research we present and analyse two in-depth case studies (Eisenhardt 1989). The two companies differ significantly with regard to their openness for collaborative open foresight.

The paper is organised into four sections. The first part explores the literature regarding corporate culture, openness and open foresight. The following section outlines the methodology – a case study approach – including information about the cases, the gained data and the methods used for analysis. The third section discusses and compares the findings with literature, and the final section contains limitations and the contribution of the paper.

Theoretical background

It has been reported that open foresight is still in its infancy (Gattringer, Wiener, and Strehl 2017) and that there exist no studies regarding cultural antecedents conducive to openness for open foresight. In contrast, the idea of openness for open innovation is increasingly discussed by innovation researchers (e.g. Herzog 2011; Laursen and Salter 2006; Lazzarotti and Manzini 2009; Van der Meer 2007). Thus we draw from insights and first findings of open innovation research.

Open foresight at the fuzzy front end of an open innovation process

Traditional corporate foresight can be characterised as a 'process of developing a range of views of possible ways in which the future could develop, and understanding these sufficiently well to be able to decide what decisions can be taken today to create the best possible tomorrow.' (Horton 1999, 5). Nevertheless, corporate foresight faces limitations: Companies are (inter alia) influenced by internal structures and restricted to existing mental models (Heger 2014; Ruff 2006). In addition, companies are often confronted with limited resources (time and budget) and with low methodological knowledge (Mietzner and Reger 2009).

These limitations led to the development of a new approach: Open foresight. Similar to the open innovation approach (Chesbrough 2003), foresight evolved into open foresight. Before now, several authors have discussed this approach (e.g. Burmeister and Schulz-Montag 2009; Daheim and Uerz 2008; Miemis, Smart, and Brigis 2012; Rau, Schweitzer, and Gassmann 2014; Ramos, Mansfield, and Priday 2012; Ruff 2006; Schmidthuber and Wiener 2017). Furthermore, first findings regarding open foresight processes (Gattringer 2016) and the role of corporate culture and partner selection in this context have been presented (Gattringer, Wiener, and Strehl 2017; Wiener, Gattringer, and Strehl 2016).

However, the definition of open foresight is very vague (Ehls et al. 2016) and the literature review shows that there are various forms of open approaches. A closer look to the research field of open innovation shows that an open innovation process is very open if the participation is open to anyone (Pisano and Verganti 2008). Crowd sourcing is an example of a totally open collaboration where everyone can join. Otherwise there are also closed forms (e.g. private clubs) with few partners (Lazzarotti and Manzini 2009).

Following this systematization, we can specify the types of open foresight as follows: crowd sourcing activities (Ehls et al. 2016; Miemis, Smart, and Brigis 2012; Ramos, Mansfield, and Priday 2012); networked foresight (foresight in innovation networks; often with many participants/companies; Heger and Boman 2015; Van der Duin, Heger, and Schlesinger 2014); participative foresight projects (various internal and external stakeholders or experts are integrated in a single corporate or public foresight project; Andersen and Andersen 2014; Heger and Rohrbeck 2012; Könnölä et al. 2009; Ramos, Mansfield, and Priday 2012), or collaborative open foresight (a closed form of inter-organisational collaborations with a small number of participating companies; Burmeister and Schulz-Montag 2009; Gattringer, Wiener, and Strehl 2017; Rohrbeck, Hölzle, and Gemünden 2009). In this study we focus on collaborative open foresight, a research field which is becoming increasingly important (Heger and Boman 2015; Rohrbeck, Battistella, and Huizingh 2015).

Based on Gattringer, Wiener, and Strehl (2017, 300) we define collaborative open foresight as:

> a discussion and analysis process of a few organizations concerning future developments in specific search fields which are relevant for the participating organizations and wherein issues related to future individual strategy and innovation options are collectively considered.

Lazzarotti and Manzini (2009) claim – in the context of open innovation – that closed forms of collaboration can not only be coordinated and controlled more easily, but can also avoid high organisational risks and costs. Furthermore, Pisano and Verganti (2008) argue that less open approaches should be preferred if strong coordination and integration is important.

Thus, collaborative open foresight can also be an effective method to initiate innovative ideas in the fuzzy front end of the innovation process for the participating companies.

The great advantages of collaborative open foresight are the comprehensive information processing and that resources (time, budget, methodological knowledge) and expertise can be shared. Another very important benefit is that through the joint work with other companies future knowledge and innovative ideas can be generated together and then used on a single company level. In this way collaborative foresight can foster 'out-of-the-box-thinking' and the risk of being limited to existing mental models can be avoided. (Gattringer, Wiener, and Strehl 2017; Heger and Boman 2015; Keinz and Prügl 2010; Van der Duin, Heger, and Schlesinger 2014).

Corporate culture and openness

Barney (1986, 657) defines corporate culture as a *'complex set of values, beliefs, assumptions and symbols that define the way in which a firm conducts its business'*. Saffold (1988) identifies shared values as core of a corporate culture, with cultural strength indicating to what extent organisation members share values. Corporate culture not only defines how companies conduct their business. According to Quinn and McGrath (1985) culture includes beliefs about organisational purpose, performance criteria, location of authority, legitimacy of power, decision-making processes, leadership style, compliance, evaluations and motivation.

The literature review shows that corporate culture is a key dimension influencing open innovation success (Docherty 2006; Enkel, Bell, and Hogenkamp 2011; Herzog and Leker 2010; Remneland-Wikhamn and Wikhamn 2011; Szymańska 2016). It is argued that an innovation-oriented culture contributes to openness for new ideas (Stock, Six, and Zacharias 2013) and it *'is an important driver of radical innovation'* (Tellis, Prabhu, and Chandy 2009, *15*). This openness – in the sense of *'a general willingness to access new knowledge, a positive attitude towards complementary influence, and, consequently, the willingness to exchange with others'* (Enkel 2010, 349) – subsequently determines an organisations' innovation performance (Laursen and Salter 2006).

Following the definition of Barney (1986) that organisational culture determines the way organisations conduct their business, we can assume that changing the approach of a company from closed to open innovation is also a matter of culture. Mortara et al. (2010, 371) argue that *'[o]pen practices have deep resonance and implication for everyone in the firm involved with innovation'* and *'needs an open culture and an open state of mind'* (Giannopouou, Yström, and Ollila 2011, 519).

In the context of inter-organizational-collaborations an unfavourable organisational culture can cause problems and can constrain the boundary-spanning activities of a firm (Van der Duin and Den Hartigh 2009; Wu, Lin, and Chen 2013). In particular the so-called 'not-invented-here-syndrome' has been highlighted as a challenge in open innovation activities (Elmquist, Fredberg, and Ollila 2009; Gassmann 2006; Keupp and Gassmann 2009). It is argued that a certain culture of resistance toward ideas from outside, scepticism and disparagement make it difficult to integrate external knowledge into the innovation process. Szymańska (2016, 151) states that this challenge of absorption of open innovation is a question of culture and describes a culture supportive of openness as follows: *'This culture must be characterised by the focus on the development, commitment and dynamism on the part of all employees as well as by being integrated with the company's objectives, being pro-innovative, open to change and cooperation with the environment.'*

The few previous findings show that in an open system a different way of thinking is required (Van der Meer 2007). Nevertheless, this issue is largely unaddressed in academia (Mortara et al. 2010).

The concepts of organisational culture are manifold and often specified too broadly in order to represent a concrete and operational basis for the analysis of the interrelationship between organisational culture and openness for innovation in our study 'Openness for the participation in collaborative open foresight'. '[M]*ost studies continue to adopt undifferentiated approaches that do not distinguish different elements of a corporate culture',* argue Stock, Six, and Zacharias (2013, 283). To analyse the culture required for open innovation, Herzog (2011) therefore recommends to conduct research on different levels e.g. corporate level or inter-organisational networks.

The focus of this research is on the corporate level. For the analysis of corporate culture itself we use the Organizational Cultural Assessment Instrument which is based on the Competing Values Framework developed by Cameron and Quinn (2006). It provides a basis to concretise 'culture' and to identify key variables influencing the interrelationship with openness for collaborative open foresight. To measure openness within the collaborative approach of open foresight, we refer to a well-established open innovation measure by Laursen and Salter (2006, 140): The *'two variables reflecting openness in terms of external search strategies'* are breadth and depth.

The Competing Values Framework is also tested and frequently applied in research (Yu 2009), especially in innovation research (e.g. Igo and Skitmore 2006; Klimas 2016) and *'provides a meaningful structure for the ideational aspects of organizational culture'* (Büschgens, Bausch, and Balkin 2013, 763). In a study review Yu (2009) argues that the Competing Values Framework is one of the most frequently used models to conduct research on organisational culture. Büschgens, Bausch, and Balkin (2013) conclude in their meta-level analysis that this framework can be used to analyse and describe the congruence of different corporate cultural profiles with organisational goals of innovation.

The Competing Values Framework was initially based on research aimed at identifying indicators and values of effectiveness within organisations (Quinn and Rohrbaugh 1983). The framework is designed along two dimensions: (1) flexibility and discretion versus stability and control, and (2) external focus and differentiation versus internal focus and integration (Cameron and Quinn 2006). Along these dimensions, four types of organisational culture are defined: clan, adhocracy, hierarchy, and market culture. Each type contains a set of effectiveness criteria, which define the core values of the respective cultural profiles (Cameron and Quinn 2006). Figure 1 shows these attributes characterising the four cultural types.

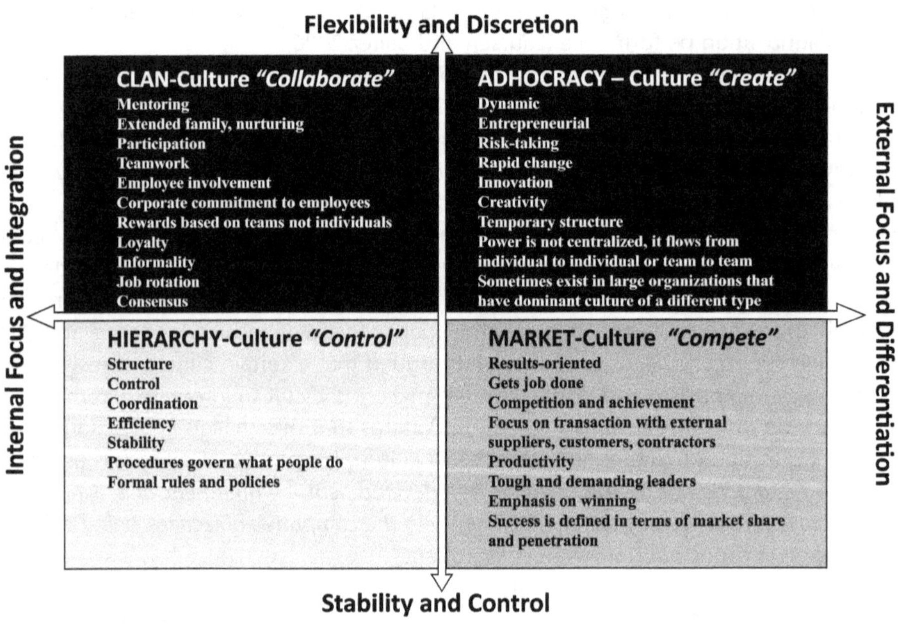

Figure 1. Characteristics of clan, adhocracy market and hierarchy types (adapted from Cameron and Quinn 2006; Wiewiora et al. 2013, 1166).

In the Competing Values Framework, organisations that are internally/control oriented and show the classical attributes of a bureaucracy are labelled **hierarchy cultures**. The hierarchy culture has a clear organisational structure, strict control, standardised rules and procedures, and well-defined responsibilities. The **clan culture** is like the hierarchy culture located along the internal/integration dimension, but in contrast focuses on flexibility and discretion and is similar to a family-type organisation. It encourages broad participation of employees, makes human resource development a priority, has an atmosphere of cohesion and mutual help, and emphasises teamwork and empowerment. The **market culture** focuses on the external rather than the internal environment, and is stability and control oriented. Market cultures are characterised by clarity of tasks and goals. An emphasis is on earning profits through both external and internal competition, efficiency and measurable outcomes. The externally/differentiation and flexibility/discretion oriented **adhocracy culture** assumes that adaption to market developments and innovativeness leads to new resources and profitability. A major goal of an adhocracy is to foster adaptability, flexibility, and creativity where uncertainty, ambiguity and information overload are typical (Cameron and Quinn 2006).

Method – case study design

The empirical analysis follows a case study (Yin 2003) approach. Yin stresses that case studies are especially suitable for the study of individuals or social units such as groups and companies. Furthermore Eisenhardt (1989) argues that case studies enable the generation of new theories. Eisenhardt (1989, 548) concludes: *'Building theory from case study research is most appropriate in the early stages of research on a topic or to provide freshness in perspective to an already researched topic.'*

Two stages of analysis are conducted: Within – and cross-case (Eisenhardt 1989). The comparison of the two cases describes and analyses if there exists a significant difference in the openness of companies participating in open foresight processes compared to companies that do not engage in open approaches. In addition, the key characteristics of corporate culture that determine this openness are of interest.

Case selection

Companies that have already participated in open foresight processes are rare. Therefore, Yin's (2003, 77) view was used as background for the case study selection: The selection can be 'straightforward': Cases (companies) are selected that are well known or enable good access to data for the researcher. In order to provide answers to the research question, *'excellent or rich examples of the phenomenon [...], but not unusual cases'* (Patton 1990, 171) are searched for. Following this 'intensity sampling' approach, two companies that differ regarding their openness for open foresight were needed.

To find these two companies, semi-structured interviews (Hopf 1978) were conducted in various companies beginning in May 2015. The selection of the two cases was due to their statements referring to the openness of the company. On the one hand, we found a company that explicitly declines open approaches as the management fears the sharing of know-how with other companies. On the other hand, we found a company which is open for open foresight processes. This company has already participated twice in collaborative open foresight projects and evaluated these participations as successful and therefore fulfilled the criteria for the case selection.

Hence, the proposed research is conducted with two companies – company 'Gamma' and 'Delta' (companies are anonymized).

Gamma and Delta are multinational companies headquartered in Austria. Both are operating in the B2B field and develop and market industrial products. In 2011/12 and again in 2014/15, Gamma participated in collaborative open foresight processes to detect and analyze trends and weak signals with regard to the future. In contrast, Delta has never participated in any open approaches. Due to the fact that the company is a world leader within its niche, the executive board is convinced that the interexchange with other companies about future trends would only mean to cede know-how and consequently to lose a competitive advantage.

Operationalisation and data collection

To meet the test of construct validity Yin (2003) recommends the use of multiple sources of evidence. This provides the opportunity to generate more compelling results and greater potential for explanation, and a stronger base for theory building can be developed. In addition, Wiewiora et al. (2013, 1171) advise the combining of quantitative and qualitative methods: *'the use of a mixed-method approach is preferred for investigating organizational context, such as culture, as it provides insightful findings with increased rigor'*. Following these suggestions, three data sources were captured to ensure adequate triangulation and to reduce bias or error in observation and field work in this study (Denzin 1970): (1) Review of organisational documentation and observation of the corporate culture and artifacts during company visits, (2) Semi-structured interviews, and (3) OCAI-questionnaire (Organizational Culture Assessment Instrument).

Artifacts can be myths and stories, but also *'attractive meeting and discussion areas in which information regarding innovations can be exchanged informally'* (Stock, Six, and Zacharias 2013, 286). Artifacts have a symbolic power and have the power to inspire (Dandridge, Mitroff, and Joyce 1980). Therefore, the researchers not only conducted interviews during their company visits, but also tried to capture artifacts. In addition, documents related to corporate culture and strategy were analysed. Gamma and Delta not only gave access to their training materials for new employees (including company presentation and core values), but they also offered internal documents about the corporate strategy and provided answers to questions that occurred.

Further, semi-structured interviews (Hopf 1978) were conducted to obtain insights into the innovation process, to inquire the factors that influence the companies' openness and to collect information needed to measure openness according to Laursen and Salter (2006). Originally the authors define 16 different sources of information for innovation, including e.g. technical standards; eight sources refer to inter-organisational collaborative innovation: *'suppliers, clients or customers, competitors, consultants, commercial laboratories/R&D enterprises, universities or other higher education institutes, government research organizations, or private research institutes'* (Laursen and Salter 2006, 140). As collaborative open foresight is not only about gaining information from external sources, but is a collaborative process (Daheim and Uerz 2008), we decided to focus on these eight sources and added another cooperation possibility: companies from other industries (see Table 1). This decision is based on a recent empirical work from Gattringer (2016), who developed an open foresight process design and reported that the open foresight participants were among others

Table 1. Breadth and depth (adapted from Laursen and Salter 2006, 139).

	Use (Breadth)	Intensity of inter-organisational collaboration (Depth)		
Knowledge source	⬇	Low	Medium	High ⇩
Suppliers				
Clients or customers				
Competitors				
Consultants				
Commercial laboratories/R&D enterprises				
Universities or other higher education institutes				
Government research organisations				
Private research institutes				
Companies from other sectors				

companies from differing industries. Each of the nine sources is binary coded, 0 in case the source is not used and 1 meaning that this is a cooperation partner. The points are simply added up so that a company has 0 in total when no external source is used and a value of nine in case all sources are used. The second variable depth depends on the degree the external source is used. The respondents can choose between no, low, medium or high degree of use. If the company uses one of the external sources to a high degree, this source is again coded with 1 and with 0 if the company reports no, low or a medium degree.

The interviews were guided by a questionnaire, recorded and then transcribed. The interviews were conducted in Gamma in May and June 2015 and in Delta in September and October 2015. All interviews lasted between 45 min and 1.5 h.

Six face-to-face semi-structured interviews (Hopf 1978) and in addition one telephone interview were conducted to provide insights into behavioural and attitudinal manifestations of the corporate culture. In Gamma and Delta the interviewees were only people that were either on the management board, or holding top-management positions in innovation or R&D related departments. In Gamma all interviewees were also part of the open foresight project team. Table 2 provides more detailed information on the interviewees.

For the quantitative corporate culture analysis, the cultural survey instrument 'Organizational Culture Assessment Instrument' (OCAI), developed by Cameron and Quinn (2006), was used. The OCAI was chosen because '[*t*]*he framework* […] *is* […] *supported by an extensive empirical literature and underlying dimensions that have a verified scholarly foundation*' (Cameron and Quinn 2006, 20). Six dimensions – dominant characteristics, organisational leadership, management of employees, organisation glue, strategic emphases and criteria of success – account for organisational culture. These dimensions are the basis for the questionnaire, which consists of six items, each item having four alternatives (A, B, C, D), which relate to a type of organisational culture. As the company language is German, a German version of the OCAI was applied. In 2012 Strack translated four out of six dimensions of the OCAI into German and a construct validation of the translated German version was realised. The remaining two dimensions were translated into German by the researchers and afterwards a native speaker did a back-translation and compared it to the original version.

Data analysis

In-depth case studies – observation & interviews

Based on the different modes (open versus closed) the two companies claim to be innovative and gain competitive advantage, we wanted to understand which corporate cultural values and artifacts foster the companies' openness for open foresight.

Case 1: gamma

Company Gamma was founded more than 70 years ago and has its headquarters in Austria. In 2015 Gamma had more than 3300 employees in over 60 countries worldwide. From its beginning Gamma focused on research and development and continuous innovation. In 2013 the strategic emphasis was on innovation and by 2015 Gamma was granted 928 patents.

Table 2. Case companies and interviewees.

Company	Role in the company	Department	Tenure 2015
Gamma	Innovation management	R&D	16 years
Gamma	Head of strategic management	Strategic management	29 years
Gamma	Project management	Strategic management	10 years
Gamma	Project manager	Product marketing – strategic management	9 years
Delta	Executive Board	Product Development Counsel	15 years
Delta	Innovation manager	Product Development Consult	2 years
Delta	Sales manager	Consulting and Planning	4 years

To emphasise innovation Gamma has already participated twice in collaborative open foresight processes. The first attempt was in 2011/12 together with a well-known customer. The idea to initiate this project arose in a top-management meeting of Gamma and its customer. This opening-up was along the supply chain and restricted to a well-known partner. Subsequently the open foresight outcome was discussed with a university department, also a well-established partnership. Referring to the openness understanding of Laursen and Salter (2006) breadth (number of external sources of knowledge) can be indicated with 2 and as there exists an intense cooperation with these partners also the depth can be indicated with 2.

In 2014/2015 Gamma participated again in an open foresight project initiated by a university institute and a mechatronics research centre. The aim was to detect and analyse trends and weak signals with regard to the future. The defined search field was 'Human-Machine Interfaces' and 'Machine-to-Machine-Communication'. The main goals for the five participating companies (three manufacturing companies from different industries, one established IT consulting firm with profound know-how in the search field and one start-up company with the intention to include the perspective of entrepreneurship) were to identify innovation opportunities and ideas for business model innovations. Following again the openness understanding of Laursen and Salter (2006) breadth counts 4 and includes the following categories: Consultants, university, companies from other industries, R&D enterprises – whereas depth remains 2 (University, R&D enterprise). This shows that over time Gamma's openness increased.

The project manager of the department 'product marketing' explains their motivation to participate in the open foresight project as follows: *'We try to be first movers and to set new standards. Therefore, it is of great importance to us to think out-of-the-box and engage with other companies. Otherwise we would restrict ourselves to existing mental models.'*

Gamma was convinced by the business value of the first open foresight. According to the head of strategic management the value was three fold: (1) It fostered 'out-of-the-box – thinking' (2) it saved resources, and (3) increased the potential of innovations as the joint discussion and analysis of future developments helped to generate new insights regarding the opportunities and risks of the identified disruptive technologies. Increasing openness by including more and unknown companies from different industries in the second open foresight was particularly beneficial due to the additional perspectives and inputs from various backgrounds and the comprehensive discussion process.

The mission of Gamma is to go off the established ruts and seek out new paths. By this the company wants to succeed where others have failed and, 'do the impossible' according to the head of the strategic management department – *'with no ifs and buts'. 'In short, while others develop gradually, we innovate in leaps and bounds by shifting the limits of what's possible'*, he continued.

The core values of Gamma are to think in the long-term, act responsibly and to place great emphasis on dealing with employees, customers and partners with respect. *'We're particularly concerned about the well-being of our employees. One of our responsibilities is to provide a healthy working environment'*, stated a person from the management board. He added: *'In 2007 we built a completely new site. It was important to us that the architecture of the new building reflects our core values.'* The design is open, bright and seems to be a friendly atmosphere, with atriums, uninterrupted open 'communication islands' and barrier-free, bright-as-day, spacious workplaces, as well as the freedom to move around.

Case 2: delta

Delta was founded about 50 years ago in Austria and has expanded rapidly and constantly since then. In 2015 the company generated revenues of around 470 million Euros and employed 2500 employees in 19 entities worldwide.

Delta emphasises the communication of its core values: Result-orientation, to think and act proactively, to be open-minded and to act responsibly. Even though a proactive approach is a core value,

Delta had no innovation department until September 2015. Only two weeks before the interview, an employee from the product development department was promoted as innovation manager and assigned to implement an innovation process within the company. Talking about the innovation and foresight activities up to the present date, the innovation manager explained that Delta operates in a conservative sector: '*We want to be proactive, but we are not proactive so far. It is the customer who has a problem and we try to find the solution'* approach is to provide only technologies meeting their high quality criteria in order to meet their customers' needs: Thoroughly tested new products – no trial and error approach. The interviews at Delta revealed that the management was very protective when it comes to know-how exchange and opening up the innovation process. They believed that keeping know-how within the company sustains its competitive advantage. '*We develop innovations in-house and as soon as we have a first draft we protect this idea by patent*', explains the innovation manager.

The impression of the research team was that to generate innovations Delta pins its expectation on internal competitiveness and goal-orientation. For example, on some office doors you could find certificates telling 'best sales team'. The sales manager sums up: '*We get binding monthly targets, which are in my opinion are quite motivating.*' The offices are designed as open offices, one for each department including an individual office for the department head. Separated only by a glass wall, the head of the department always has an overview of the staff. To sum up, the observation of Delta's culture appears to emphasise achievement, control, demanding leaders and competition.

OCAI-D survey results

In Gamma the OCAI-D (German version) instrument was administered to 70 randomly selected managers (sampling by personnel number) in June 2015. The executive board fully supported the survey and sent the questionnaire to the selected managers. 60 persons (=85.81%) completed the OCAI-D. The sample is composed of middle managers, upper-middle managers and the executive board. The data were collected using an online-tool. The responses were voluntary and respondents were promised confidentiality.

At the end of November 2015 the survey was conducted in the same way in Delta. The questionnaire was administered to 84 randomly selected managers and 51 persons (=60.71%) completed the OCAI-D.

Analysing the OCAI-D, descriptive results show that by tendency, the two cases differ regarding to the four cultures of the Competing Values Framework (Table 3; Figure 2).

In the case of Gamma – a company that is open for open foresight – clan values Mean (M): 32.21 (Standard Deviation (SD): 10.86) are dominant. The clan-culture score is followed by the adhocracy culture M: 25.43 (SD: 7.51), hierarchy culture scored M: 23.88 (SD: 8.90) and market culture reached the lowest score with M: 18.49 (SD: 6.53).

The culture survey in Delta shows that market values dominate with M: 33.80 (SD: 11.60), followed by the hierarchy culture M: 27.40 (SD: 7.79), clan culture scored M: 20.24 (SD: 8.16) and adhocracy culture reached the lowest score with M: 18.55 (SD: 7.09).

To test if there is a significant difference between the two cases, an independent samples t-test (two means t-test) was conducted. As independent variable (grouping variable) the two groups:

Table 3. Descriptive results.

	Gamma Mean (SD)	Delta Mean (SD)
Clan	32.21 (10.86)	20.24 (8.16)
Adhocracy	25.43 (7.51)	18.55 (7.09)
Market	18.49 (6.53)	33.80 (11.6)
Hierarchy	23.88 (8.90)	27.40 (7.79)

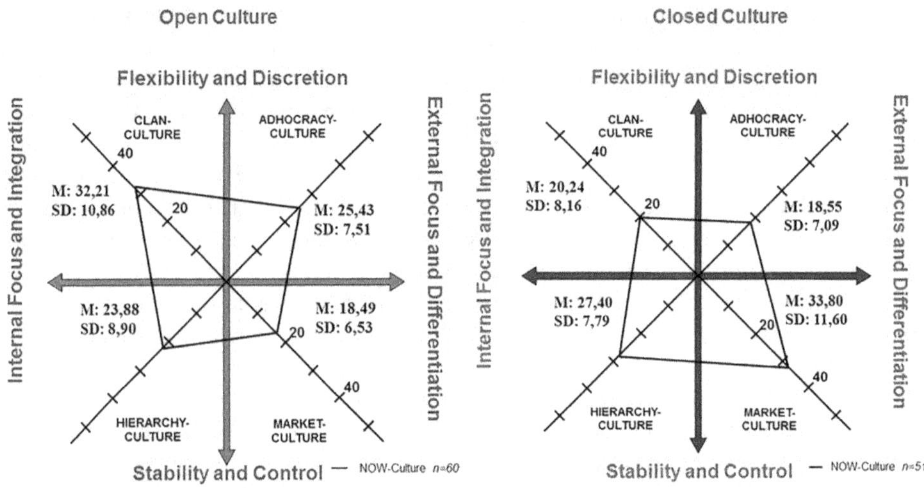

Figure 2. Open culture versus closed culture.

Gamma and Delta were defined. The cultures – clan, adhocracy, market and hierarchy – were defined as test variables (dependent variables). To test the homogeneity of variances Levene's test (Levene 1960) was used.

Clan-culture

For clan, the variances were equal for Gamma and Delta, $F(1, 108) = 3.562$; $p = .062$. On average, employees in Gamma scored higher for clan values than the employees in Delta. This difference was significant*[1] $t(108) = 6.474$, *Cohen's* $d = 1.25$, $p < .001$.

Adhocracy-culture

For adhocracy, the variances were equal for Gamma and Delta, $F(1, 108) = 0.022$; $p = .882$. On average employees in Gamma scored higher for adhocracy values than the employees from Delta. This difference was significant* $t(108) = 4.888$, *Cohen's* $d = 0.94$, $p < .001$.

Market-culture

For market, the variances were equal for Gamma and Delta, $F(1, 108) = 3.720$; $p = .056$. On average employees in Gamma scored lower for market values than the employees from Delta. This difference was significant* $t(108) = -8.715$, *Cohen's* $d = 1.68$, $p < .001$.

Hierarchy-culture

For hierarchy, the variances were equal for Gamma and Delta $F(1, 108) = 1.347$; $p = .248$. On average employees in Gamma scored lower for hierarchy values than the employees from Delta. This difference was significant* $t(108) = -2.166$, *Cohen's* $d = 0.42$, $p = .033$.

Discussion

Following Cameron and Quinn (2006) the quadrant in which the score is highest indicates the predominant cultural dimension of the company. It represents the basic assumptions and values that predominate. In the case of Gamma a culture that is characterised by flexibility/discretion (clan & adhocracy) values appears to support openness for participation in collaborative open foresight. Similar to the empirical findings, various authors also point out the importance of flexibility oriented cultural values (e.g. Stock, Six, and Zacharias 2013; Biazzo 2009; Buganza and Verganti 2006) and openness to innovation (Laursen and Salter 2006).

A culture that is dominated by clan has strong loyalty and traditional values. Leaders are seen as mentors and the place of work seems to be like an extended family (Cameron and Quinn 2006). Further supportiveness is a value in the clan quadrant and fosters employees' propensity to suggest new ideas, as they experience a feeling of psychological safety (Büschgens, Bausch, and Balkin 2013). In the literature it is argued that a culture characterised by an open climate (Remneland-Wikhamn and Wikhamn 2011) is important. Naqshbandi, Kaur, and Ma (2015) identified in their study of Malaysian high-tech companies organisational culture types that enable open innovation activities. The results show that a highly integrative culture enables in-bound open innovation. Tsui, Wang, and Xin (2006, 359) describe this culture type as: '*a culture profile characterized by high scores on both internal integration and external adaptation*' (similar to the 'clan' culture in the Cameron and Freeman (1991) study).

Adhocracy, the second quadrant located along the flexibility/discretion dimension, is described as a dynamic, entrepreneurial and creative work environment, where managers are considered to take risks, show commitment and are experimental and innovative (Cameron and Quinn 2006). The external/differentiation focus of adhocracy facilitates the retrieval of information that fosters the creation of ideas, the identification of opportunities and detection of the (technological) frontier (Büschgens, Bausch, and Balkin 2013; Dyer, Gregersen, and Christensen 2008). A literature comparison of different open innovation contributions shows a similar picture. An innovation-friendly culture is described by encouraging risk-taking, supporting openness to new ideas and tolerating failure (Docherty 2006; Herzog and Leker 2010; Ollila and Elmquist 2011). Chesbrough (2003) points out the importance of the external orientation and the entrepreneurial spirit (Enkel, Bell, and Hogenkamp 2011) that is represented by an adhocracy culture. Thus, we propose:

> **P1:** A corporate culture with a tendency towards flexibility and discretion and characterized by a dominant clan culture and at the same time a distinct adhocracy culture fosters openness for collaborative open foresight participation.

With regard to innovative culture Docherty (2006) identifies key factors that should be taken under consideration: a focus on learning and not primarily on results is critical to effective innovation and a tolerance for risk taking. Especially the statement to not focus primarily on results contrasts with the core values that dominate market culture organisations, namely competitiveness and productivity. The culture survey in Delta shows that these market culture values dominate followed by hierarchy-culture. In contrast, clan culture and adhocracy culture scored lowest. It is interesting to note that even though the market culture exhibits – like the adhocracy culture – an external focus, our findings show that market-culture does not foster openness for participation in collaborative open foresight processes. In this regard Chesbrough (2003) points out one of the closed innovation principles. It states that closed companies believe in control of the innovation process and do not open it, to ensure competitors don't profit from their ideas. The executive board of Delta shares this opinion that to interexchange with other companies about future trends would only mean to cede know-how and consequently to lose the competitive advantage.

Delta, the closed company, is embossed by hierarchy values which can also be seen in the organisation chart. Naqshbandi, Kaur, and Ma (2015) emphasise that companies characterised by a hierarchy culture do not stress the values which are important for internal integration and external adaption and are therefore not fostering openness for open innovation. We therefore propose:

> **P2:** A corporate culture with a tendency towards stability and control and characterized by a dominant market culture and at the same time a distinct hierarchy culture inhibits openness for collaborative open foresight participation.

Contribution

Opening up companies' boundaries to look into the long term future jointly with other organisations and to share knowledge on possible future developments, is proposed as a promising approach to foster collection of broader data, sharing of expertise and resources, inclusion of new perspectives from different backgrounds and 'out-of-the-box-thinking' (Burmeister and Schulz-Montag 2009; Gattringer, Wiener, and Strehl 2017; Heger and Boman 2015; Keinz and Prügl 2010; Van der Duin, Heger, and Schlesinger 2014).

However, before committing to inter-organisational collaborations like collaborative open foresight, it is important to analyze if the company possesses the necessary prerequisites. Findings from adjacent areas indicate that organisational culture and its influence on a company's openness to engage in open approaches is an important precondition. For example, in the context of open innovation it is argued that corporate culture is a key dimension influencing open innovation success (Docherty 2006; Enkel, Bell, and Hogenkamp 2011; Herzog and Leker 2010; Remneland-Wikhamn and Wikhamn 2011; Szymańska 2016). In inter-organisational-collaborations an unfavourable organisational culture can cause problems and can constrain the boundary-spanning activities of a firm (Van de Vrande et al. 2009; Wu, Lin, and Chen 2013).

Based on these insights we assume that an appropriate organisational culture is also an important precondition for openness for collaborative open foresight. Therefore the following research question was investigated: *Which features of corporate culture foster or inhibit a company's openness for collaborative open foresight?*

Based on two in-depth case studies applying the Competing Values Framework for (organisational) culture analysis the following cultural values that foster or inhibit openness have been identified: (1) A corporate culture characterised by a clan culture and adhocracy culture (flexibility/discretion related values) fosters openness, whereas (2) a market and hierarchy culture inhibits (stability/control related values) openness for collaborative open foresight participation.

As research on open foresight in general is still rare and – to our knowledge – research reporting the influence of corporate culture is virtually non-existent – so far, we argue that the results of this study provide new theoretical and managerial insights for the planning and realisation of collaborative open foresight projects. Based on the findings of this research work companies can assess if their corporate culture is conducive or not for participation in collaborative open foresight processes. Taking this into consideration companies can decide if an open foresight process would be beneficial for them or not.

Limitations of the study and suggestions for further research

We suggest that further research is required to improve the understanding of corporate cultural features which inhibit or foster openness for open foresight. The following limitations of the present research work may serve as a useful starting point for follow-up studies. First, as open foresight is still in its infancy the case study approach helps to develop a rich understanding of the researched companies, but at the same time limits a generalisation of the findings to other organisations. Therefore, future studies that capture additional contextual influences on open foresight are needed. Obvious factors to be considered include the following. First, the case study only included Austrian companies and participants – are the findings applicable beyond the Austrian borders? Thus, a multinational approach is needed in future research. Second, the companies involved are relatively large – what about open foresight conducted by small and medium-sized companies? Third, openness is not tested for industry-type. So what is the role of industry-type regarding openness for collaborative open foresight? Then, a greater sample to test the generated proposition is needed. Such research would help substantially to define and specify the scope of the statements generated by the present study.

Notwithstanding the limitations discussed above, the present study provides new theoretical and empirical insights into the influence of corporate culture on openness for collaborative open foresight. We conclude that values along the flexibility/discretion dimension (clan & adhocracy) foster openness, while values along the stability/control dimension (market & hierarchy) inhibit openness.

Note

1. $^{*}\alpha = .05$.

Disclosure statement

No potential conflict of interest was reported by the authors.

References

Andersen, A. D., and P. D. Andersen. 2014. "Innovation System Foresight." *Technological Forecasting and Social Change* 88: 276–286.

Barney, J. B. 1986. "Organizational Culture: Can It Be a Source of Sustained Competitive Advantage?" *The Academy of Management Review* 11 (3): 656–665.

Biazzo, S. 2009. "Flexibility, Structuration, and Simultaneity in New Product Development." *Journal of Productive Innovation Management* 26 (3): 336–353.

Boer, H., and F. Gertsen. 2003. "From Continuous Improvement to Continuous Innovation: A (Retro)(per)Spective." *International Journal of Technology Management* 26 (8): 805–827.

Brettel, M., and N. J. Cleven. 2011. "Innovation Culture, Collaboration with External Partners and NPD Performance." *Creativity and Innovation Management* 20 (4): 253–272.

Buganza, T., and R. Verganti. 2006. "Life-Cycle Flexibility: How to Measure and Improve the Innovative Capability in Turbulent Environments." *Journal of Productive Innovation Management* 23 (5): 393–399.

Burmeister, K., and B. Schulz-Montag. 2009. "Corporate Foresight: Praxis und Perspektiven." In *Zukunftsforschung und Zukunftsgestaltung: Beiträge aus Wissenschaft und Praxis*, edited by Reinhold Popp and Elmar Schüll, 277–921. Berlin: Springer.

Büschgens, T., A. Bausch, and D. B. Balkin. 2013. "Organizational Culture and Innovation: A Meta Analytic Review." *Journal of Productive Innovation Management* 30 (4): 763–781.

Cameron, K. S., and S. J. Freeman. 1991. "Cultural Congruence, Strength, and Type: Relationships to Effectiveness." *Research in Organizational Change and Development* 5: 23–58.

Cameron, K. S., and R. E. Quinn. 2006. *Diagnosing and Changing Organizational Culture: Based on the Competing Values Framework*. San Francisco: Jossey-Bass.

Chesbrough, H. W. 2003. *Open Innovation: The new Imperative for Creating and Profiting From Technology*. Boston: Harvard Business School.

Daheim, C., and G. Uerz. 2008. "Corporate Foresight in Europe: From Trend Based Logics to Open Foresight." *Technology Analysis & Strategic Management* 20 (3): 321–336.

Dandridge, T. C., I. Mitroff, and W. F. Joyce. 1980. "Organizational Symbolism: A Topic to Expand Organizational Analysis." *Academy of Management Review* 5 (1): 77–82.

Denzin, N. K. 1970. *The Research Act: A Theoretical Introduction to Sociological Methods*. 2nd ed. Chicago: AldinePublishing Company.

Docherty, M. 2006. "Primer on Open Innovation: Principles and Practice." *PDMA Visions* 30 (2): 13–17.

Dyer, J. H., H. B. Gregersen, and C. M. Christensen. 2008. "Entrepreneur Behaviors, Opportunity Recognition, and the Origins of Innovative Ventures." *Strategic Entrepreneurship Journal* 2 (4): 317–338.

Ehls, D., S. Korreck, R. Jahn, M. A. Zeng, S. Heuschneider, C. Herstatt, H. Koller, and S. Spaeth. 2016. "Open Foresight: Exploiting Information from External Sources." *SSRN*. https://ssrn.com/abstract = 2764208

Eisenhardt, K. M. 1989. "Building Theories From Case Study Research." *Academy of Management Review* 14 (4): 532–550.

Elmquist, M., T. Fredberg, and S. Ollila. 2009. "Exploring the Field of Open Innovation." *European Journal of Innovation Management* 12 (3): 326–345.

Enkel, E. 2010. "Attributes Required for Profiting From Open Innovation in Networks." *International Journal of Technology Management* 52 (3/4): 344–371.

Enkel, E., J. Bell, and H. Hogenkamp. 2011. "Open Innovation Maturity Framework." *International Journal of Innovation Management* 15 (06): 1161–1189.

Fink, M., R. Lang, and R. Harms. 2013. "Local Responses to Global Technological Change — Contrasting Restructuring Practices in two Rural Communities in Austria." *Technological Forecasting and Social Change* 80 (2): 243–252.

Gassmann, O. 2006. "Opening up the Innovation Process: Towards an Agenda." *R&D Management* 36: 223–228.

Gattringer, R. 2016. "Open Foresight-Prozesse: Eine Action Research Studie zur Identifikation von Schlüsselaktivitäten und -Faktoren bei der Planung und Umsetzung von Open Foresight-Prozessen." *Habilitationsschrift*, Johannes Kepler Universität.

Gattringer, R., M. Wiener, and F. Strehl. 2017. "The Challenge of Partner Selection in Collaborative Foresight Projects." *Technological Forecasting and Social Change* 120: 298–310.

Giannopouou, E., A. Yström, and S. Ollila. 2011. "Turning Open Innovation Into Practice: Open Innovation Research Through the lens of Managers." *International Journal of Innovation Management* 15 (3): 505–524.

Hansen, E. G., A. C. Bullinger, and R. Reichwald. 2011. "Sustainability Innovation Contests: Evaluating Contributions with an eco Impact-Innovativeness Typology." *International Journal of Innovation and Sustainable Development* 5 (2/3): 221–245.

Hansen, E. G., and J. Klewitz. 2012. "The Role of an SME's Green Strategy in Public-Private Eco-Innovation Initiatives: The Case of Ecoprofit." *Journal of Small Business & Entrepreneurship* 25 (4): 451–477.

Heger, T. 2014. "A Theoretical Model for Networked Foresight." The XXV ISPIM Conference – Innovation for Sustainable Economy & Society, Dublin, June 8–11.

Heger, T., and M. Boman. 2015. "Networked Foresight—the Case of EIT ICT Labs." *Technological Forecasting and Social Change* 101: 147–164.

Heger, T., and R. Rohrbeck. 2012. "Strategic Foresight for Collaborative Exploration of New Business Fields." *Technological Forecasting and Social Change* 79 (5): 819–831.

Herzog, P. 2011. *Open and Closed Innovation: Different Cultures for Different Strategies*. Wiesbaden: Gabler Verlag.

Herzog, P., and J. Leker. 2010. "Open and Closed Innovation - Different Innovation Cultures for Different Strategies." *International Journal of Technology Management* 52 (3/4): 322–343.

Hopf, C. 1978. "Die Pseudo-Exploration: Überlegungen zur Technik Qualitativer Interviews in der Sozialforschung." *Zeitschrift für Soziologie* 7 (2): 97–115.

Horton, A. 1999. "Forefront: a Simple Guide to Successful Foresight." *Foresight - The Journal of Future Studies, Strategic Thinking and Policy* 1 (1): 5–9.

Huang, L., Y. Guo, A. L. Porter, J. Youtie, and D. K. R. Robinson. 2012. "Visualising Potential Innovation Pathways in a Workshop Setting: the Case of Nano-Enabled Biosensors." *Technology Analysis & Strategic Management* 24 (5): 527–542.

Igo, T., and M. Skitmore. 2006. "Diagnosing the Organizational Culture of an Australian Engineering Consultancy Using the Competing Values Framework." *Construction Innovation* 6 (2): 121–139.

Keinz, P., and R. Prügl. 2010. "A User Community-Based Approach to Leveraging Technological Competences: An Exploratory Case Study of a Technology Start-Up From MIT." *Creativity and Innovation Management* 19 (3): 269–289.

Keupp, M. M., and O. Gassmann. 2009. "Determinants and Archetype Users of Open Innovation." *R&D Management* 39 (4): 331–341.

Klimas, P. 2016. "Organizational Culture and Coopetition: An Exploratory Study of the Features, Models and Role in the Polish Aviation Industry." *Industrial Marketing Management* 53: 91–102.

Könnölä, T., T. Ahlqvist, A. Eerola, S. Kivisaari, and R. Koivisto. 2009. "Management of Foresight Portfolio: Analysis of Modular Foresight Projects at Contract Research Organisation." *Technology Analysis & Strategic Management* 21 (3): 381–405.

Lahi, A., and T. Elenurm. 2015. "SME Open Innovation Implicating Factors in Different Innovation Phases." *International Journal of Management Science and Information Technology* 16: 29–45.

Laursen, K., and A. Salter. 2006. "Open for Innovation: The Role of Openness in Explaining Innovation Performance among U.K. Manufacturing Firms." *Strategic Management Journal* 27 (2): 131–150.

Lazzarotti, V., and R. Manzini. 2009. "Different Modes of Open Innovation: A Theoretical Framework and an Empirical Study." *International Journal of Innovation Management* 13 (04): 615–636.

Levene, H. 1960. "Robust Tests for Equality of Variances." In *Contributions to Probability and Statistics*, edited by Ingram Olkin, 278–292. Stanford: Stanford University Press.

Miemis, V., J. Smart, and A. Brigis. 2012. "Open Foresight." *Journal of Future Studies* 17 (1): 91–98.

Mietzner, D., and G. Reger. 2009. "Practices of Strategic Foresight in Biotech Companies." *International Journal of Innovation Management* 13: 273–294.

Mortara, L., I. Slacik, J. J. Napp, and T. Minshall. 2010. "Implementing Open Innovation: Cultural Issues." *International Journal of Entrepreneurship and Innovation Management* 11 (4): 369 397.

Naqshbandi, M. M., Kaur, S., and Ma, P. 2015. "What Organizational Culture Types Enable and Retard Open Innovation?." *Quality & Quantity* 49 (5): 2123–2144.

Ollila, S., and M. Elmquist. 2011. "Managing Open Innovation: Exploring Challenges at the Interfaces of an Open Innovation Arena." *Creativity and Innovation Management* 20 (4): 273–283.

Paliokaitė, A., N. Pačėsa, and D. Sarpong. 2014. "Conceptualizing Strategic Foresight: An Integrated Framework." *Strategic Change* 23 (3-4): 161–169.

Patton, M. Q. 1990. *Qualitative Evaluation and Research Methods*. Newbury Park: Sage Publications.

Pisano, G., and R. Verganti. 2008. "Which Kind of Collaboration is Right for you?" *Harvard Business Review* 86 (12): 78–86.

Tellis, G. J., J. C. Prabhu, and R. K. Chandy. 2009. "Radical Innovation Across Nations: The Preeminence of Corporate Culture." *Journal of Marketing* 73 (1): 3–23.

Quinn, R. E., and M. R. McGrath. 1985. "The Transformation of Organizational Cultures: A Competing Values Perspective." In *Organizational Culture*, edited by Peter J. Frost, Larry F. Moore, Meryl R. Louis, Craig C. Lundberg and Joanne Martin, 315–334. Beverly Hills: Sage Publications.

Quinn, R. E., and J. Rohrbaugh. 1983. "A Spatial Model of Effectiveness Criteria: Towards a Competing Values Approach to Organizational Analysis." *Management Science* 29 (3): 363–377.

Ramos, J., T. Mansfield, and G. Priday. 2012. "Foresight in a Network Era: Peer-Producing Alternative Futures." *Journal of Future Studies* 17 (1): 71–90.

Rau, C., Schweitzer, F., and Gassmann, O. 2014. "Open Foresight Workshops for Opportunity Identification." In *Open Innovation: New Product Development Essentials From the PDMA*, edited by Charles H. Noble, Serdar S. Durmusoglu, and Abbie Griffin, 27–52. Hoboken, NJ, USA: Wiley.

Remneland-Wikhamn, B., and W. Wikhamn. 2011. "Open Innovation Climate Measure: The Introduction of a Validated Scale." *Creativity and Innovation Management* 20 (4): 284–295.

Rohrbeck, R. 2014. "Trend Scanning, Scouting and Foresight-Techniques." In *Management of the Fuzzy Front End of Innovation*, edited by Oliver Gassmann and Fiona Schweitzer, 59–74. Cham: Springer International Publishing.

Rohrbeck, R., C. Battistella, and E. Huizingh. 2015. "Corporate Foresight: An Emerging Field with a Rich Tradition." *Technological Forecasting and Social Change* 101: 1–9.

Rohrbeck, R., K. Hölzle, and H. G. Gemünden. 2009. "Opening up for Competitive Advantage: How Deutsche Telekom Creates an Open Innovation Ecosystem." *R&D Management* 39 (4): 420–430.

Ruff, F. 2006. "Corporate Foresight: Integrating the Future Business Environment Into Innovation and Strategy." *International Journal of Technology Management* 34 (3-4): 278–295.

Saffold, G. S. 1988. "Culture Traits, Strength, and Organizational Performance: Moving Beyond "Strong" Culture." *Academy of Management Review* 13 (4): 546–558.

Sarpong, D., M. Maclean, and C. Davies. 2013. "A Matter of Foresight: How Practices Enable (or Impede) Organizational Foresightfulness." *European Management Journal* 31 (6): 613–625.

Schmidthuber, L., and M. Wiener. 2017. "Aiming for a Sustainable Future: Conceptualizing Public Open Foresight." *Public Management Review*, March 1. https://scholar.google.com/citations?view_op=view_citation&hl=en&user=b5l9xLEAAAAJ&citation_for_view=b5l9xLEAAAAJ:zYLM7Y9cAGgC.

Stock, R. M., B. Six, and N. A. Zacharias. 2013. "Linking Multiple Layers of Innovation-Oriented Corporate Culture, Product Program Innovativeness, and Business Performance: A Contingency Approach." *Journal of the Academy of Marketing Science* 41 (3): 283–299.

Strack, M. 2012. "Organisationskultur im Competing Values Model: Messeigenschaften der Deutschen Adaption des OCAI." *Journal of Business and Media Psychology* 3 (1): 30 41.

Szymańska, K. 2016. "Organisational Culture as a Part in the Development of Open Innovation the Perspective of Small and medium-Sized Enterprises." *Management* 20 (1): 142–154

Tsui, A. S., H. Wang, and K. R. Xin. 2006. "Organizational Culture in China: An Analysis of Culture Dimensions and Culture Types." *Management and Organization Review* 2 (3): 345–376.

Van de Vrande, V., J. P. J. de Jong, W. Vanhaverbeke, and M. de Rochemont. 2009. "Open Innovation in SMEs: Trends, Motives and Management Challenges." *Technovation* 29 (6-7). 423–437.

Van der Duin, P. A., and E. Den Hartigh. 2009. "Keeping the Balance: Exploring the Link of Futures Research with Innovation and Strategy Processes." *Technology Analysis & Strategic Management* 21 (3): 333–351.

Van der Duin, P. A., T. Heger, and M. D. Schlesinger. 2014. "Towards Networked Foresight? Exploring the use of Futures Research in Innovation Networks." *Futures* 59: 62–78.

Van der Meer, H. 2007. "Open Innovation: The Dutch Treat: Challenges in Thinking in Business Models." *Creativity and Innovation Management* 16 (2): 192–202.

Vecchiato, R., and C. Roveda. 2010. "Strategic Foresight in Corporate Organizations: Handling the Effect and Response Uncertainty of Technology and Social Drivers of Change." *Technological Forecasting and Social Change* 9 (77): 1527–1539.

Weber, K. M., J. C. Harper, T. Konnola, and V. Carabias Barcelo. 2012. "Coping with a Fast-Changing World: Towards new Systems of Future-Oriented Technology Analysis." *Science and Public Policy* 39: 153–165.

Wiener, M., Gattringer, R., and Strehl, F. 2016. "Corporate Cultural Features Conducive For Participation In Open-Foresight." Strategic Management Conference: Strategy Challenges in the 21st Century: Innovation, Entrepreneurship and Coopetition, Rome, June 5–7.

Wiewiora, A., B. Trigunarsyah, G. Murphy, and V. Coffey. 2013. "Organizational Culture and Willingness to Share Knowledge: A Competing Values Perspective in Australian Context." *International Journal of Project Management* 31 (8): 1163–1174.

Wu, Y.-C., B.-W. Lin, and C.-J. Chen. 2013. "How Do Internal Openness and External Openness Affect Innovation Capabilities and Firm Performance?" *IEEE Transactions on Engineering Management* 60 (4): 704–716.

Yin, R. K. 2003. *Case Study Research: Design and Methods*. Applied Social Research Methods Series v. 5. Thousand Oaks: Sage Publications.

Yu, T. 2009. "A Review of Study on the Competing Values Framework." *International Journal of Business and Management* 4 (7): 37–42.

Potentials of collaborative foresight for SMEs

Yulia Milshina and Konstantin Vishnevskiy

ABSTRACT
Fast speed of changes induces all types of business enterprises, including small and medium enterprises (SMEs), to seek for instruments of innovation development. Foresight process is one of the most efficient methods of strategic planning that allows setting goals and estimating precisely the potential consequences of developing new technologies, products, and services. Taking into account SMEs features (resource limitations, scope and time horizon of foresight process etc.), while building innovation strategy and conducting corporate foresight, we propose collaborative foresight that allows overcoming these obstacles. This process has a synergetic effect not only on SMEs and its competitiveness, but also on region as a whole.

Introduction

In rapidly changing environment any innovative business needs the best available information to form its development strategy and to optimise its technological plans. The more dynamic firm are it has more demand for foresight knowledge in order to secure its investment decisions and lower risks.

Empirical corporate research foresight processes are widespread in large corporation all over the world, but the foresight needs and activities of SMEs stays below the radar. However, the value of SMEs' contribution to employment, value creation, and innovation is well recognised (Jannek and Burmeister 2007). Bidaurratzaga and Dell (2012) argue that SMEs represent the basis of most economies, but often the innovation process within small companies lacks systemic view as in foresight process and limited by market or technology analysis.

Operational processes dominate and SME almost have no chance to create the strategic vision for the future, so studies almost all relate to a specific project work and they haven't got a systemic view. (Van de Vrande, Jong, and Vanhaverbeke 2009). The innovation strategy building of large companies is thoroughly studied in academic literature, but methodological approaches for small and medium firmsare much less covered (Poteralska and Sacio-Szymańska 2014; Bidaurratzaga and Dell 2012; Hideg, Nováky, and Alács 2014; Paliokaite 2010). Today resources (human, financial etc.) are restricted both in large and small businesses, but for SMEs these restrictions become a significant barrier for innovation development and foresight implementation.

One of the features of SMEs is that their technological ideas often appear as more innovative and risky than those of large firms (Audretsch and Vivarelli 1996; Lee et al. 2010). However, large firms tend to conduct their innovation explorations systematically (Lichtenthaler et al. 2004), and these processes in SMEs is less systematic. Furthermore, most of SMEs lack diverse portfolios or plentiful innovation resources, so technology opportunities have seldom been defined and analysed (Savioz and Blum 2002; Cho et al. 2016).

SMEs, which use foresight to achieve the strategic management process, may follow one or combine two of the following organisational approaches (Cuhls and Johnston 2006):

(1) Internal foresight implementation.
(2) Usage of external foresight activities performed by other actors.

SMEs foresight activities and roles in the regional and technological development are very different in literature. Grim offered Foresight Maturity Model, which consists of six disciplines defining the best practices for foresight field (leadership, framing, scanning, forecasting, visioning, planning), provides numerical assessment of current practices, and can be adopted to SMEs practices (Grim 2009). The successful development of this model is presented in Kononiuk and Sacio-Szymańska research (Kononiuk and Sacio-Szymańska 2015).

Rohrbeck and Gemünden (2011) identified three roles of corporate foresight by using a four-step innovation process as a frame of reference: strategist role (at the start of innovation process – creating a vision, providing strategic guidance, consolidating opinions, assessing and repositioning innovation portfolios, and identifying the new business models of competitors), initiator role (outside the innovation process – identifying new customer needs, technologies, and product concepts of competitors), opponent role (along the innovation process – challenging basic assumptions, challenging the state-of-the-art of current R&D projects, and scanning for disruptions that could endanger current and future innovations).

Saritas develops an idea of the Systemic Foresight Methodology (SFM) that proposes a framework for implementing foresight. STM assumes the analytical work with complex interplay between the Social, Technological, Economic, Ecological, Political and Value (STEEPV) systems. There are a series of iterative phases (Saritas 2013): intelligence, imagination, integration, interpretation, intervention, impact, interaction phase.

Resource constraints do not allow SMEs companies to advance innovation and explore new market opportunities. By using collaborative foresight efforts SMEs can exploit cluster potentials and activate innovation capabilities. Foresight support system (FSS) which concept was presented by Keller et al. assumes collaborative usage of foresight instruments electronically to support a continuous foresight process. The authors argue that clustering is a much-cited approach for resource-constrained firms to innovate in difficult circumstances of increased competition and market dynamics. Applying foresight collaboratively within cluster all participants (large, small and medium companies, research organisations and governmental institutions) can be prepared to overcome the challenges of discontinuous change (Keller, Markmann, and Heiko 2015).

Markmann et al. offer Competitive monitor (CoMo) project as a joint research project aiming to create foresight platform within the cluster provided a foresight toolbox, that can be used by SME to collaboratively generate, discuss, evaluate and develop future knowledge in the cluster (Markmann et al. 2012).

The aim of our research is to define the role of collaborative foresight as the platform where different by size firms unites their efforts and where SME can overcome the limitation of implementing foresight. Collective action by small firms in rural and urban communities, clusters, and cooperatives helped SMEs overcome their size disadvantage (Sato 2015; Markmann et al. 2012).

The following research questions are stated:

- What are the differences of foresight processes in large companies and SMEs?
- Which methods and how successfully are used for carrying out the foresight in SMEs?
- How can SMEs use the potential of a collaborative foresight and united efforts of different stakeholders for foresight implementation?

The structure of the paper is following. First section is devoted to the features of SMEs while implementing foresight and to the crucial role of the united efforts and intermediary. The next section presents the overview of foresight processes in large companies and SMEs. The following section is focused on analysis of existing foresight research methods for SMEs. The next part is about role of associations (on the example of cluster) for conducting foresight collaboratively in SMEs. The outcomes and benefits of collaborative foresight implementation are distinguished on the basis of the Russian case study in the last section.

Foresight processes in large companies and SMEs

Existing papers on foresight use terms such as corporate foresight (Rohrbeck 2010; Vishnevskiy, Karasev, and Meissner 2015b), strategic foresight (Vishnevskiy, Meissner, and Karasev 2015a; Sarpong, Maclean, and Alexander 2013). These notions describe research activities of future in various types of organisations (Kononiuk and Sacio-Szymańska 2015). Vishnevskiy et al. argue that strategic foresight allows the assessment of a company's prospects and takes into account of a variety of parameters which makes it one of the most effective tools for strategic decision-making (Vishnevskiy, Meissner, and Karasev 2015a). Other researchers (Vecchiato and Roveda 2010) apply strategic foresight to emphasise the existing correlation between foresight and strategy building.

The innovation process within SMEs is often partially described and implemented, because operational activities frequently take the most time and SMEs find it hard to create strategic visions for the future (Van de Vrande, Jong, and Vanhaverbeke 2009). Innovation promoting methodological approaches should be in constant correlation with features of SME.

Stakeholders (large companies, government, scientific community representatives) play an important role of strategy development for SMEs due to their interest of SMEs innovative development.

All firms (small and large) face the necessity to form innovative development strategies and to optimise future technology plans (Vishnevskiy and Karasev 2015). Some scholars devote their research to comparative analysis between large corporations and SMEs while corporate foresight implementing. (Bidaurratzaga and Dell 2012; Jun, Seo, and Son 2013). The dissimilarity between the scope of activities, forms of ownership, and availability of resources in large companies and SMEs explains the difference (Table 1).

The history of foresight implementation in large companies has counted several decades and it has been thoroughly studied by many researchers and experts who worked out numerous methods of foresight research (Becker 2002; Popper Villarroel and Popper 2008). The general character of corporate foresight for large companies is multifunctionality and consistency. Large firms are more likely to have sufficient resources at their disposal to implement strategic foresight, and also be more likely to want a return on their investment in it (Rohrbeck and Schwarz 2013). Usually several organisational departments are involved in foresight research resulted often in fundamentally new knowledge creation (Slaughter 1999). The longevity and wide scope of foresight then require substantial resource spending and justified for middle or long term period of planning. As opposite, foresight projects prepared by SMEs is usually situational and forced by urgent needs in particular cases to adapt existing knowledge to create new products (Bidaurratzaga and Dell 2012). Efforts of foresight implementation are often low effective because of strict resource limits in SMEs, and planning horizon is short (up to 1 or 2 years with some exceptions).

The purposes for foresight implementation in large corporations and SMEs are nearly identical. Any company seeks for development strategy in order to effectively organise and design the innovation policies, to choose the relevant products and new markets (Slaughter 1999; Horton 1999; Becker 2002; Daheim and Uerz 2008; Gracht, Vennemann, and Darcow 2010; Fikirkoca and Saritas 2012; Battistella, De Toni, and Pillon 2015).

Large firms are also time-bounded, so in the course of a single foresight corporations apply 5–6 tools for studying potential future by combining qualitative, quantitative, synthetic methods and creating priority systems, from which they form roadmaps and development scenarios (Becker

Table 1. Difference between foresight processes in large companies and SMEs

Comparison criteria	Foresight for SMEs	Corporate Foresight for large companies
General features	• Situational character • Spot analysis • SWOT-Analysis • Existing knowledge adaptation	• Multifunctionality (involves several departments of the organisation) • Consistency • Systematic character • New knowledge creation
Purpose	• Make obvious strategy to find investors and appropriate partners • Adaptation to changing environment conditions • Potential risks identifying • New product launches	• Innovation strategies formation • Adaptation to changing environment conditions • Potential risks identifying • New product launches • New market development
Horizon of planning	Often short-term (up to 1–2 years), sometimes for longer period	Mid- or long term (3–5 years or longer)
Own resources for foresight implementation	All sorts of resources are limited or some types of resources may be in abundance while others lack	In sufficient quantity – financial and human resources In limited quantity – time resources
Barriers for foresight	• Lack of financial, human and time resources • Much of the time is spending on operational activity • Lack of incentives to the anticipation of the future • The risk of development	• Non-serious attitude of top management and stakeholders to future information application • Intraorganizational hierarchy • The existing bonus, career and control systems • Frequent personnel changes among top management • Lack of resources

Source: authors.

2002; Popper Villarroel and Popper 2008). Defining prospective product line is most common for SMEs, therefore foresight for SMEs is often confined to roadmap development for communication with investors (Holmes and Ferrill 2005; Jun, Seo, and Son 2013). At the initial stage of strategic system development external stakeholders (government, large companies, and scientific community) who are interested in the innovative development of SMEs can play an important role in conducting foresight studies for SMEs.

Significant barriers for foresight process in SMEs are following: lack of financial, human and time resources, lack of incentives to the anticipation of the future, risks of development in competitive environment. For large firms the barriers are pretty much the same (lack of resources is not so critical, but nevertheless is important) with some particular differences. Large firms have much more possibilities to hedge a lot of risks by their diversified product/technology portfolio. Often success of the foresight project depends on attitude of the top management and key stakeholders to future information application.

It should be mentioned that strong connection exists between the strategy, foresight, and organisational culture in any company, including SMEs (Kononiuk and Sacio-Szymańska 2015). Culture basically as proved by (Rohrbeck and Gemünden 2011) influences significantly the innovation process. According to (Rohrbeck et al. 2009), several factors can be defined that have an impact on the success of foresight. These are the following:

- information usage;
- method sophistication;
- people and networks;
- organisation of foresight activities within companies;
- corporate culture.

Despite the fact that there are many similarities between foresight in large companies and SMEs, there are many differences arising mainly from the nature of these types of companies. Large firms generally have more sufficient resources and more diversified structure to implement foresight. SMEs concentrate on the operational issues and have strict resource limitations so they are recommended to act collaboratively during the foresight implementation with the support of intermediary.

Methodological approaches for foresight in SMEs

According to Poteralska and Sacio-Szymańska (2014), corporate foresights for SMEs can be divided into three groups: foresights conducted for and by enterprises; foresight for the needs of an enterprise, for example, with the help of external non-profit association or company (Dutch organisation – Syntens NT); foresight projects, which are not executed by enterprises, but the enterprises participate in them or are their beneficiaries (national or regional foresight efforts) (Keenan, Marvin, and Winters 2002).

There are slightly many projects concerning foresight application in SMEs particularly in technology-based ones. Jannek and Burmeister (2007) argue that complex foresight researches are conducted significantly rarely . The staff constraints and unavailability of additional funds constitute severe limitations to applicate more complicated foresights. But scenario-based methods are quite regularly applied (Andriopoulos and Gotsi 2006).

Roadmaps become widespread as they help SMEs identifying new technologies and products in order to win or strengthen the position on the market (Holmes and Ferrill 2005; Jun, Seo, and Son 2013). Roadmaps as a methodology which is frequently used in foresight in different ways (Vishnevskiy et al., 2015; Karasev and Vishnevsky 2010; Kindras et al. 2014) have many benefits both for SMEs and large companies. Roadmaps represent flexible instrument which can be applied for technology and innovation development, strategy forming, human resource management etc. Roadmapping allows actors, SMEs as well, to develop a path along which future research and development should be continued and to identify useful partners (private and public) to cooperate. Klein emphasises that the role of government is critical for support the issue of trust among participants in collaborative foresightprocess. The use of corporate foresight opportunities by SMEs is also probable (Paliokaite 2010).

The foresight initiatives on general SMEs-related issues is often organised or supported by different state or over- or international agencies such as EC, EACEA, ACEAN etc. For example, ASEAN 'Strategic Action Plan for SME Development 2016–2025' executes the vision through the mission statement, strategic goals and related desired outcomes and actions (ASEAN 2015). Action paths in timelines are realised through The Implementation Roadmap, which is also the basis for the monitoring mechanisms to manage the progress of the action lines on regular basis. Key performance indicators (KPIs) provide policy indicators to evaluate the achievement of the strategic goals (Figure 1).

One of the leading countries in SME development is Singapore. SPRING focuses on developing clusters and strengthening the environment for enterprise development. These plans are based on foresight instruments: trend monitoring, scanning, roadmapping etc. Roadmapping was one of the key methods that were used already in 2005 for SME development (Holmes and Ferrill 2005).

The objective of the SME Technology Roadmapping Programme, implemented in South Korea at the federal level, was to enhance the ability of SMEs to plan R&D activity (Jun, Seo, and Son 2013). Most of SMEsbenefited from the roadmapping and eventually developed the strategic technologies indicated in the roadmap. South Korea example is comparable with the Singapore programme in creating operational and technology roadmaps, however in the Korean case the programme worked with smaller companies who required the involvement of more financial and human resources.

Hideg, Nováky, and Alács (2014) gave a priority to the scenario building for SME. Among the scenario-building stakeholder groups, two characteristic groups were formed, visioning the region's futures. They compare the economic scenarios of SMEs and social scenarios of the 'Hungary 2025' research project (Nováky 2010) because the relation between the two-type groups of stakeholders can represent the relation of the part and the whole.

Figure 1. Structure of Strategic Action Plan for SME. Source: ASEAN 2015.

The concept of Opportunity Landscape created by (Savioz and Blum 2002) represents an accessible system of administrative support for SMEs and is based on the Technology Intelligence method and some strategic approaches. The purpose of Opportunity Landscape is the identification and anticipation of future trends in scientific environment by means of their systematic observation and with the help the necessary technological information for decision-making.

Pozdnakova (2008) paid closer attention to Knowledge Management, which is the process of the systematic creation, preservation, distribution and application of the elements of organisational intellectual capital. Knowledge Management involves such aspects as strategy, technology, enterprise resources, organisational culture, business processes, and the processes of knowledge processing. The final model contains common elements of the SME environment, Knowledge Management and foresight.

Bidaurratzaga and Dell (2012) proposed the Future Garage Process which represents a specially designed foresight for SMEs and based on an analysis of the literature, expert pools and case studies. Future Garage Process brings weak signals, leading indicators and long-term impact factors in SME innovative strategies and helps to overcome some common barriers for SMEs.Summing up, we distinguish several approaches for SME foresight: Roadmaping, Opportunity Landscape, Knowledge Management, Future Garage Process, and Collaborative Foresight. In our preliminary investigation, we formulated methodological approaches to foresight research for SMEs (Vishnevskiy and Egorova 2015). The most popular foresight methods used by SMEs are literature review, expert pools,, scenario building, brainstorming, expert interviews, desk research, market analysis, technology research (Table 2). We have analysed many literature sources and found out the most popular approaches among researchers and added a list of selected methods.

The affordability of described approaches should be disclosed during the first stages of particular foresight project because both the objectives and the organisation of these approaches are significantly different and the majority of these methods have not yet been widely studied in practice. The specific choice must be based on the objectives of the proposed research, the available resources, and the readiness of SMEs to implement such approaches.

Based on the analysis of these approaches we applied collaborative foresight concept in our case study based on foresight in cluster. In business and innovation clusters the environment facilitates networking and cooperation among companies and institutions that contributes to organisational learning and enabling even SMEs to profit from economies of scale and scope while staying flexible (Porter 1998; Cooke 2001). Collaboration of SMEs while conducting foresight expand the scope of future awareness and peripheral vision and may facilitate the SMEs in clusters in implementing

Table 2. Methodological approaches to foresight in SMEs and selected methods.

Methodological approaches to foresight in SMEs	Literature sources (examples)	Selected methods
Scenario building	• Hideg, Nováky, and Alács (2014) • Andriopoulos and Gotsi (2006) • Amankwah-Amoah and Sarpong (2016)	• Literature review • Futures workshops • Evaluation and comparative analysis of scenarios
Opportunity landscape	• Savioz and Blum (2002) • Arshed, Bunduchi, and Finch (2012) • Poteralska and Sacio-Szymańska (2014)	• Top-down/bottom-up approaches • Expert analysis methods • Market and technology analysis
Knowledge management	• Pozdnakova (2008) • Bušinska, Stecjuka, and Kirikova (2008)	• Brainstorming • Competitor analysis • Knowledge audit • teamwork
Collaborative foresight	• Markmann et al. (2012) • Heger and Boman (2015)	• Markets estimates • Bibliometric analysis • Future seminars
Future Garage Process	• Bidaurratzaga and Dell (2012)	• Bibliometric analysis and literature review • Expert polls • Case studies • Weak signals
Roadmapping	• Holmes and Ferrill (2005) • Jun et al.(2013) • Battistella, De Toni, and Pillon (2015)	• Expert studies • Brainstorming • Market and technology analysis • Scenarios

Source: Authors.

research and innovation and dealing not only with incremental, but discontinuous change (Keller, Markmann, and Heiko 2015).

Role of clusters for conducting collaborative foresight in SMEs

The concept of clusters has been widely adopted in many countries all over the world in recent decades (OECD 2007; Pro Inno Europe 2012; Oxford Research 2008). The demand for this concept can be explained by opportunities of effective cooperation between actors, better access to innovation, technologies, infrastructure, specific services, expert executives, growth of competitiveness and quality of products and services improvement (Kutsenko 2015).

Supplying competitive products to the global market, clusters become growth points, influencing regional and national economies (Abashkin, Boyarov, and Kutsenko 2012). Conducting foresight researches for determining and forming future is much more spread within European innovative territorial clusters. It is connected with the higher popularity of regional Foresight in Europe (Keenan, Marvin, and Winters 2002; Kindras et al. 2014).

Formation and development of territorial clusters is an effective mechanism for building successful firm-to-firm cooperation (Sölvell and Williams 2013). The development of clusters stimulates the increase in labour productivity, the formation of new companies and the creation of new jobs, promotes the growth of the innovative potential of the territories, the formation of competitive advantages and the unique image (brand) of the region or territory that facilitates attracting investment in regions.

The specificity of the cluster approach to the development of the territory lies in the complexity and systemic nature of setting the tasks for the development of the territory, enhancing synergetic effects from the use of various instruments of regional development. To promote the cluster efficiency a continuous foresight process should be applied. Engaging in foresight (Keller, Markmann, and Heiko 2015; OECD 2007):

- enables clusters and organisations to face discontinuous change;
- avoids lock-in of business clusters by networking and knowledge exploitation;
- collaborative foresight can support cluster networking.

Cooperation of SMEs with other firms within cluster offers them: more channels for learning and creating experience; economies of scale; economies of scope; higher flexibility and shared risks. SMEs are usually the main policy target for encouraging inter-firms networking and involving large companies in these networks influence positively on SMEs' capabilities as applicants of new innovation and market opportunities (Boekholt and Thuriaux 1999).

This interaction between members of cluster including SMEs during the collaborative foresight implies a combination of internal and relational resources, which adds an additional multiplicative effect on the firm's performance due to the synergy obtained (Hervas-Oliver and Albors-Garrigos 2008).

Success regions should adopt foresight in order to identify and develop their resources, carrying on collaborative processes to plan and implement change and growth strategies (Battistella and Pillon 2016). While large firms develop constantly foresight and technology roadmaps, SMEs could benefit from a foresight studies at regional level (Bassani, Minola, and Vismara 2016). Some researchers find strong evidence that cluster policy brings additional positive effect to existing policy of SMEs which are the main actors in the cluster development process, in relation to whether their performance has been improved as a result of cluster effects (Karaev, Koh, and Szamosi 2007). So this increase in competitiveness could be strongly supported by foresight implementation in SME. Collaborative, continuous, and action-oriented foresight in clusters implemented by iterative bottom-up processes create a possibility for companies to transfer their innovation agenda into action (Salmenkaita and Salo 2004). Successful clusters that experience free flow of information and ideas are best suited for collaborative foresight. Foresight is observed by policymakers as one of the main activities for achievement of long term and even short term goals, and topical example is the ROADMAP 2020+ project. The whole complicated process of strategy-making has a name STRATCLU (Strategic Cluster Development) (Clar and Sautter 2014). The learning cycle had three main stages: stock-taking: review of cluster position in the global context; forward-looking: longer-term perspectives and priorities; action-planning: roadmaps with milestones and specific joint actions (roadmapping).

The significant place of the whole cluster and the Foresight process is given to SMEs. The MicroTEC Sűdewest approach is seen as best practice example for 'enterpreneurial process' recommended by the European Commission to develop regional smart specialisation strategies and to benefit on internal and external markets.

Technology intelligence methodology for Technology Roadmapping (TRM) tailored for SMEs was developed by Battistella et al. It was created for the S&T Park in Italy, Area Science Park, which became an innovation intermediary and had a need of supporting its activities of SMEs' technology transfer services with a theoretical methodology for foresight. The research strategy was based on action research with two cycles of action. The Opportunity profile is aimed at simplifying the traditional methodology of TRM. The Extended Map relies on collaboration of SMEs coordinated by an intermediary and presents an ecosystemic view to the methodology (Batistella 2014).

Collaboration between SMEs seems to be very effective in foresight preparation and realisation context. Cooperation allows reducing the necessity for resources and to stimulate learning and networking activities. In order to capitalise knowledge and financial resources, there should be an involvement of several SMEs existing under the cluster or SMEs' association roof and they should have interests in the same industry or technological area. The cluster can play an important role as an innovation intermediary for the project coordination, methodological support, provision of foresight tools and the access to information sources.

Summing up, the major outcomes of using the potential of cluster to organise and support foresight process in SMEs might be divided in 2 groups:

- straight: increase of participants, innovative products and enhancement of cooperation etc.;
- indirect: rise of the growth rates of industrial production in the region, GRP and GDP increase, investment flow to the region, etc.

Benefits from clusters for SMEs foresight research (the Russian case study)

Turning attention to the Russian Federation, nowadays a huge number of different clusters are planned and already created in such areas as Moscow, Rostov, Altay, Samara, etc. The system of cluster development included into the system of innovation strategy in Russia consists of industries and regional strategies, development strategies of universities and 'anchor' leading companies of cluster as well as SMEs support programme (Kutsenko 2015).

Methodology of roadmapping

The approved methodological approach (Vishnevskiy et al. 2015a; Karasev and Vishnevskiy 2013; Vishnevskiy, Karasev, and Meissner 2015b) was applied for this case study and modified for the purposes of innovation development and foresight implementation in SMEs. Studying of international best practice and practical experience in Russia allows authors to propose own approach to corporate foresight. Foresight process consists of scenarios creation of long-term scientific, technological and innovative development determined by long-term trends and challenges to identify 'points of growth'.

The process of developing the roadmap includes 5 phases (Figure 2):

(1) Pre-roadmapping. The project domain and key priorities are defined. Delphi method and expert polls are used also.
(2) Desk research. All available information and knowledge are gathered. For this purpose, a literature review that provides initial information about prospective technologies and products is undertaken. Using bibliometric and patent analysis, research fronts are identified and a list of prospective innovation related decisions for a subject field. Scanning process is aimed at external factors studying to create a list of risks.
(3) Expert procedures. Expert methods occupy a special place and lead to the discovery of 'tacit knowledge'. They include a number of expert interviews with representatives of business, academia, and public authorities.
(4) Creative analysis. The roadmapping team and selected experts analyze all data collected in previous stages. SWOT-analysis reveals the key potentials and obstacles of the development of the subject field. Backcasting means the identification of most attractive future visions. This phase results in a set of possible vectors of innovation development, reflecting the whole technological chain, which involves R&D – technology – product – market. The core element of the integrated roadmap approach combining both technology push and market pull is a cross-impact analysis (Vishnevskiy, Karasev, and Meissner 2015b).
(5) Interactive discussion. On this stage, a scenario approach is used to decrease uncertainty about the future. This stage includes workshops in order to discuss possible scenarios of subject field development.

Foresight methods assessment was conducted during the survey (more than 20 experts). The involvement of experts allowed us to obtain integral estimates of the frequently used methods and to classify them by their cost and effectiveness.

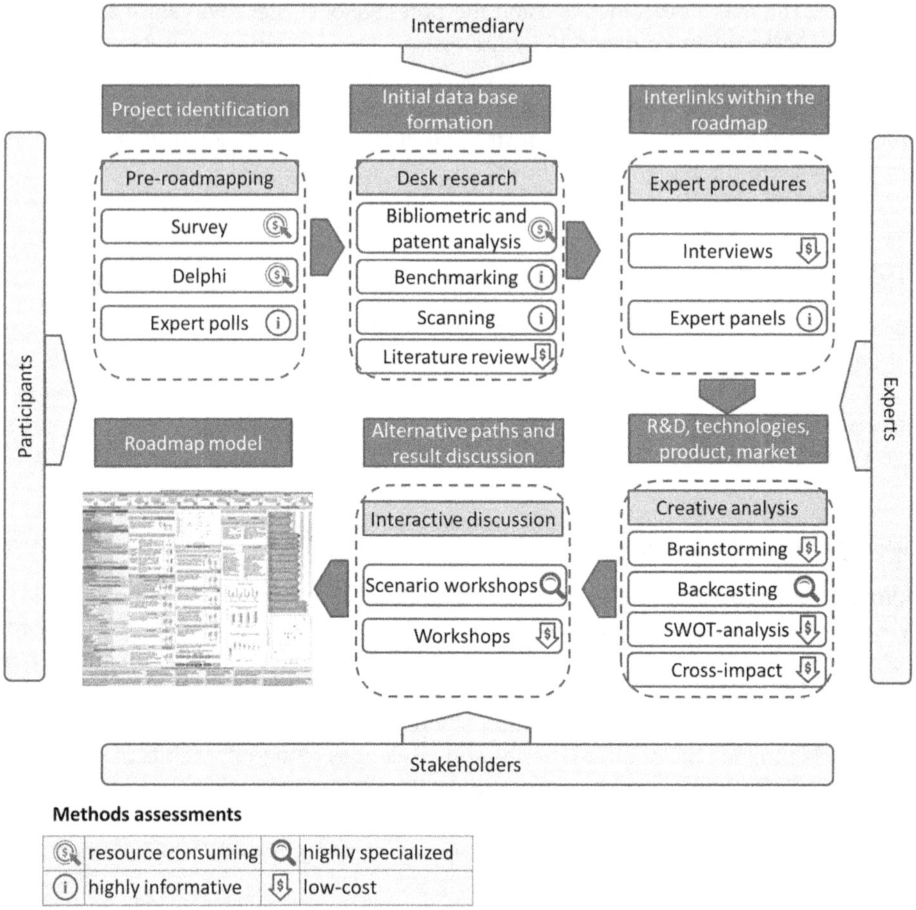

Figure 2. Scheme of proposed integrated roadmap approach. Source: authors.

As for the external circle of the roadmapping process, it includes following subjects: intermediary, experts, participant and other stakeholders. The role of intermediary is very important and, as noticed by Battistella, De Toni, and Pillon (2015), its activity is not just limited by 'the traditional role of facilitating'. While foresight is implemented in cluster collaboratively, the intermediary supports the group of SMEs and it is responsible for project management, providing tools and skills of intelligence and facilitates access to knowledge resources.

Supporting industrial and cluster innovation strategies, roadmaps may trace the further steps of innovation development by providing a set of specific action plans in several areas: R&D, sales and marketing, risks and challenges, key stakeholders, implementation of advanced manufacturing technologies, development of innovation products. These plans allow the beneficiaries companies within cluster to identify the long-term goals, to launch innovation products that will be in high demand.

A case study: foresight for SMEs in the medical cluster

The case study is based on the research that has been performed during the preparation of roadmap for The Medical Cluster (the name is disguised for confidentiality reasons). The Cluster comprised key organisations and companies active in medical biotechnology, production of medical equipment, and provision of high-technology medical services. During the preparation of the development project for the Cluster we have piloted the methodological approach of collaborative foresight including roadmap construction for the whole cluster, part of which was the roadmap for SMEs included in the cluster. This allows us to translate innovation priorities and indicators, investigate alternative scenarios and build the future perspectives for a number of small and medium-sized innovative companies. The scheme of this integrated roadmap with several layers is on the figure below (Figure 3).

The roadmap consists of the following layers: mission, priorities and targets; measures of support; scenarios of development; key performance indicators (KPI); key stakeholders. In each of these layers, there was a part devoted to SME development within the Cluster. Main priorities and targets for SME development were indicated and classified for short, mid- and long-term. Measures of support for SME were developed according to defined priorities and targets, each supportive measure related to one or several priorities and targets for their realisation.

Summing up, comprehensive methodology for a successful implementation of foresight in SMEs is needed to be investigated and covered. Most of the existing approaches have been developed for a specific project aimed at the development of the SME. We have modified our 5-step foresight methodology of Roadmapping for SMEs based on cost-benefit analysis of each foresight method. This creates a system of priorities for the development of SMEs, and offers an action plan for their implementation.

Results and discussion

Cooperation in cluster contributes to forming vital business ecosystem because it supports the idea of interacting organisations and individuals that coevolve and share common resources. This process solves the problem of resource limitation and lack of knowledge. Uniting the efforts of SMEs within a cluster supports the innovation spreading and implementation. Harmaakorpi and Uotila (2006) argue that only foresight embedded with the region's stakeholders can overcome the 'black holes of regional strategy making'.

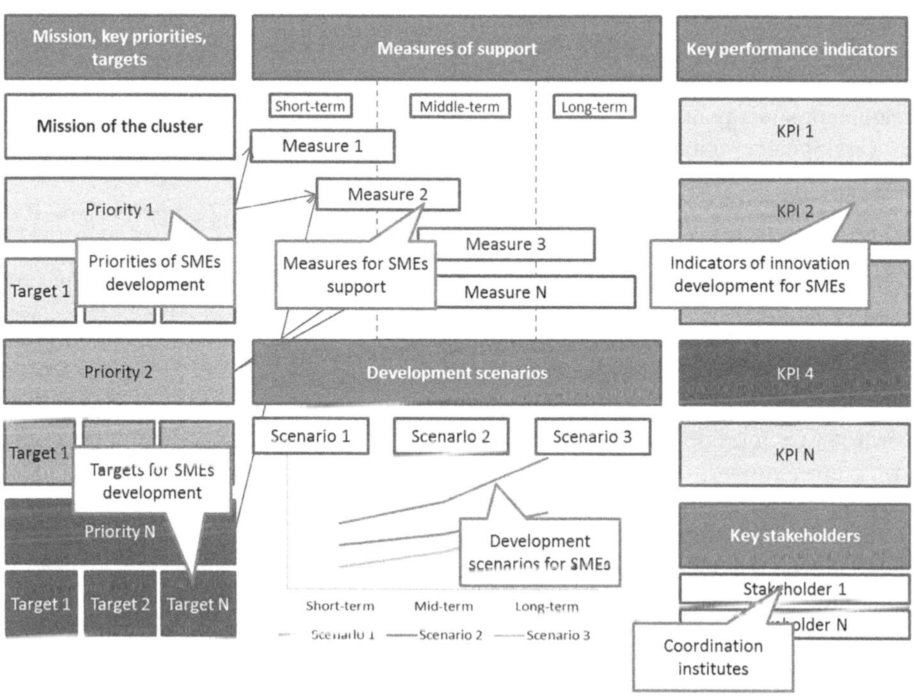

Figure 3. Foresight for SMEs as a part of roadmap for the cluster. Source: authors.

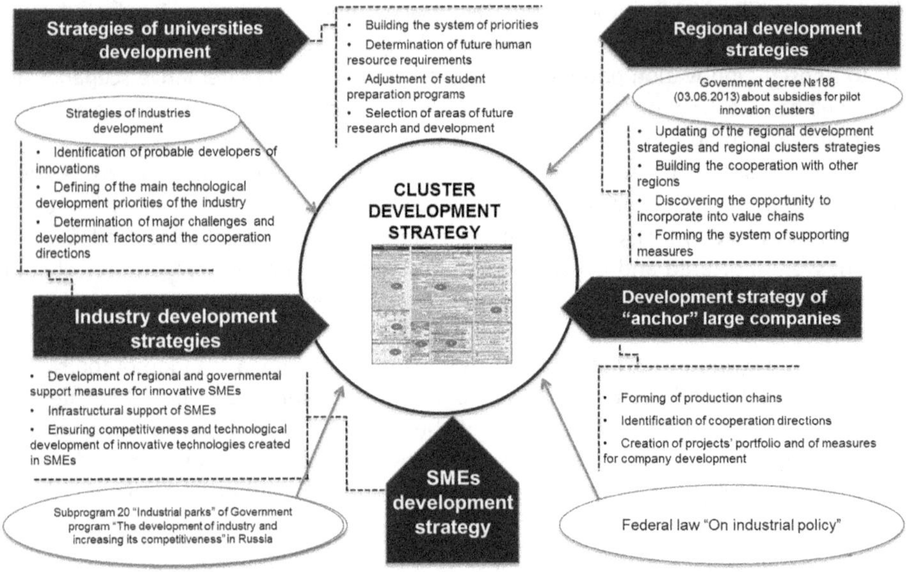

Figure 4. SME development initiative included into the cluster development programme. Source: authors.

Suggested approach contributes not only to SMEs innovation and market development but also to regional development strategy. The pilot project approved that foresight process for SMEs can be successfully inscribed into the cluster strategy development which takes its important place in the system of strategic planning in Russia (Figure 4).

The Cluster development strategy included several interlinked trajectories that have their own strategy development plan:

- Strategies of development of universities.
- Development strategies of the region.
- Industry development strategy.
- Development strategy of 'anchor' (leading) large companies.
- SMEs development strategy.

The role of innovation intermediary, cluster administration, is very important because it helps to define the objectives of Foresight co-research within a group of SMEs; supply the tools and competences, enable the access to the knowledge sources; manage project and coordinates the companies group and collaboration activities.

We assessed the major effects of cluster development programme and carrying out foresight collaboratively using PEST analysis (Figure 5).

Cluster's actors (companies, research organisations and governmental institutions) work collaboratively to apply foresight and they can be well prepared for future changes and can benefit in fast moving environment.

Collaborative foresight became the basis of SMEs fostering competitiveness and helped 6 of them to build strategic partnership with large companies within cluster. 9 start-ups created by universities occupied their niches in the cluster value chain of innovative products. The following major projects with participation of SMEs, universities and large companies were initialised:

- creation of biomedical cellular products for regenerative medicine (implementation period: 4 years). Effects: reducing the cost of treatment by decrease in the length of stay of patients in the hospital, reducing the cost of drug therapy, payments for temporary working incapacity.

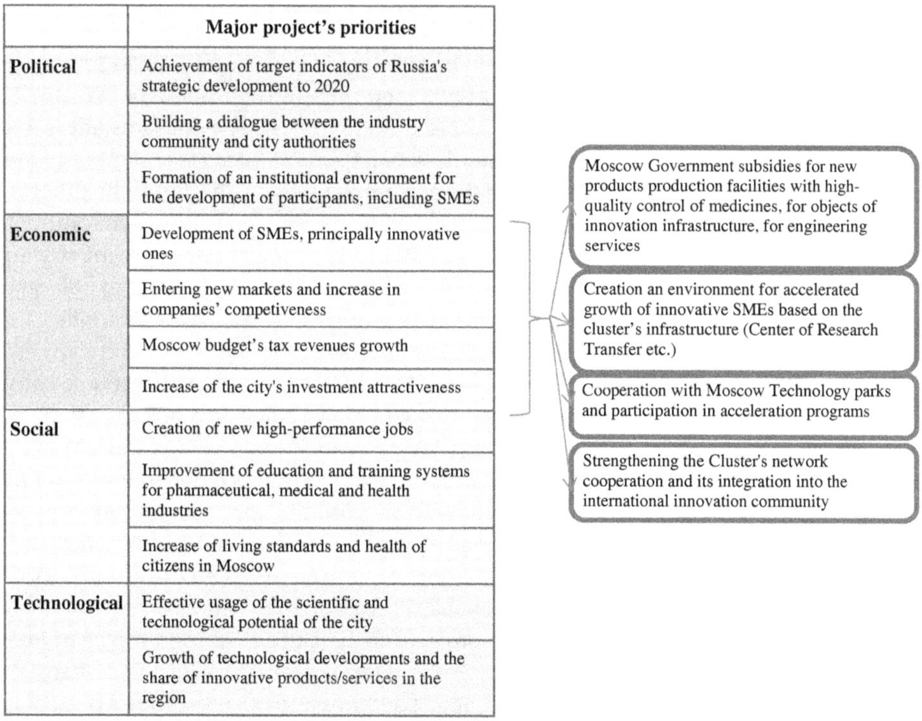

Figure 5. Identification of major project's priorities and practical steps to achieve them by using PEST-analysis. Source: authors.

- noninvasive diagnostics of oncological diseases (implementation period: 5 years). Effects: decrease in the level of oncological diseases; reduction of budgetary expenses for treatment and improvement of quality and duration of active life of patients with oncological diseases.
- diagnostic test system, which allows to analyze the most important oncological mutations in human regulatory genes in tumour DNA (implementation period: 5 years). Effect: reduce the cost of drug therapy, payments for temporary incapacity for work.

Implementation of collaborative foresight into the cluster contributed to an 15% increase in the number of medical SMEs that were created with the support of local authorities (with the participation of graduates of Moscow universities and scientific organisations). Evaluation conducted a year ago showed that performance indicators of SMEs (turnover, employment) which have participated in collaborative foresight were much better than those which have not.

There are still number of questions for further research and discussion such as:

(1) How to initiate and stabilise the cooperation between SMEs as well as multinational companies, government, universities.
(2) Profound research of obstacles for foresight implementation within SMEs in clusters and associations is needed.
(3) More detailed view on the role of Intermediary in SMEs cooperation and innovation process.
(4) The choice of relevant methodological solution of Foresight Implementation for SMEs in clusters etc.

Conclusion

Collaborative approach allows foresight studies to be conducted in resource-limited settings and can be used by small companies and in regional foresight, long-term future studies for industrial clusters and technology platforms, and all other areas where quick and cost-efficient foresight is in a great demand. The role of uniting facility such a cluster is very important because it allows accumulating the resources and competences of different SMEs and implementing the foresight process simultaneously in these companies. Also our pilot project showed that our approach contributes not only to SMEs innovation and market development, but also to regional development strategy.

Collaborative foresight can be applied successfully in cluster. Clusters as complex self-organizing structures cannot develop without a long-term strategy and specific, achievable priorities. Development priorities based on foresight methodology give an idea of technology chains and allow establishing efficient cooperation links. SMEs often locate in clusters in order to cope with resource constraints. Companies in successful clusters readily exchange ideas and have a free flow of information. As a systematic, collaborative, and action-oriented process (Gavigan and Scapolo 2001), foresight can support an efficient exchange among actors in cluster in order to expand research and innovation services.

One of the most important questions arises: how to stabilise the cooperation between small and medium-sized companies as well as multinational companies, government, universities and so on. The cluster administration played a role of the intermediary and coordinated the group of companies and activities to be done in collaboration and enabled access to knowledge.

During the preparation of the development project for the cluster we used modified integrated roadmap methodology with the separate part for SMEs in the cluster. This allows us to translate innovation priorities and indicators, investigate alternative scenarios and build the future perspectives for a number of SMEs.

The main limitations of the proposed approach can be divided into 2 parts: limitations of integrated roadmapping methodology (the quality of information achieved from experts) and those connected with cluster approach (financial, legal, trust between members). To eliminate the first one, we use both qualitative and quantitative methods. The other limitation could be solved by enhancing and developing of trust between cluster participants with the help of intermediary.

The methodology can be developed further in terms of the set of criteria, the portfolio of foresight methods, fields for application, the groups of beneficiaries, and linkages with STI policy and funding as well as in foresight assessment of issues described in the paper.

Acknowledgements

The paper was prepared within the framework of the Basic Research Programme at the National Research University Higher School of Economics (HSE) and supported within the framework of a subsidy by the Russian Academic Excellence Project '5–100'.

Disclosure statement

No potential conflict of interest was reported by the authors.

Funding

This work was supported by The chapter was prepared within the framework of the Basic Research Program at the National Research University Higher School of Economics and supported within the framework of a subsidy by the Russian Academic Excellence Project '5-100'.

ORCID

Yulia Milshina http://orcid.org/0000-0002-7603-2602
Konstantin Vishnevskiy http://orcid.org/0000-0003-3621-0504

References

Abashkin, V., A. Boyarov, and E. Kutsenko. 2012. "Cluster Policy in Russia: From Theory to Practice." *Foresight-Russia* 6 (3): 16–27.

Amankwah-Amoah, J., and D. Sarpong. 2016. "Historical Pathways to a Green Economy: The Evolution and Scaling-up of Solar PV in Ghana, 1980–2010." *Technological Forecasting and Social Change* 102: 90–101.

Andriopoulos, C., and M. Gotsi. 2006. "Probing the Future: Mobilising Foresight in Multiple-Product Innovation Firms." *Futures* 38 (1): 50–66.

Arshed, N., R. Bunduchi, and J. Finch. 2012. "*Technology Roadmapping and SMEs: A Literature Review.*" Proceedings of druid conference, Copenhagen, Denmark.

ASEAN. 2015. *ASEAN Strategic Action Plan for SME Development 2016–2025*. Jakarta: ASEAN Secretariat.

Audretsch, D., and M. Vivarelli. 1996. "Firm Size and R&D Spillovers: Evidence From Italy." *Small Business Economics* 8: 249–258.

Bassani, G., T. Minola, and S. Vismara. 2016. "Technology Foresight for Regional Economies: A How-to-Do Guide." In *University Evolution, Entrepreneurial Activity and Regional Competitiveness*, 385–392. Switzerland: Springer International Publishing.

Battistella, C. 2014. "The Organisation of Corporate Foresight: A Multiple Case Study in the Telecommunication Industry." *Technological Forecasting and Social Change* 87: 60–79.

Battistella, C., A. F. De Toni, and R. Pillon. 2015. "The Extended Map Methodology: Technology Roadmapping for SMES Clusters." *Journal of Engineering and Technology Management* 38: 1–23.

Battistella, C., and R. Pillon. 2016. "Foresight for Regional Policy: Technological and Regional Fit." *Foresight (los Angeles, Calif)* 18 (2): 93–116.

Becker, P. 2002. *Corporate Foresight in Europe: A First Overview*. Working paper. Brussels: European Commission.

Bidaurratzaga, E., and M. Dell. 2012. "*Strategic Foresight in SMEs: Challenges and Solutions-the Future Garage Process.*" Ispim conference proceedings, 1, The International Society for Professional Innovation Management (ISPIM).

Boekholt, P., and B. Thuriaux. 1999. "Public Policies to Facilitate Clusters: Background, Rationale and Policy Practices in International Perspective." *Boosting Innovation: the Cluster Approach* 381–412.

Bušinska, L., J. Stecjuka, and M. Kirikova. 2008. "On the Structure Mirroring Approach: Foresight Module in the Knowledge Management Systems of SMES." *Applied computer systems* 34: 113–125.

Cho, C., B. Yoon, B.-Y. Coh, and S. Lee. 2016. "An Empirical Analysis on Purposes, Drivers and Activities of Technology Opportunity Discovery: The Case of Korean SMEs in the Manufacturing Sector." *R&D Management* 46: 13–35.

Clar, G., and B. Sautter. 2014. "Research Driven Clusters at the Heart of (Trans-) Regional Learning and Priority-Setting Processes." *Journal of the Knowledge Economy* 5 (1): 156–180.

Cooke, P. 2001. "Regional Innovation Systems, Clusters, and the Knowledge Economy." *Industrial and Corporate Change* 10 (4): 945–974.

Cuhls, K., and R. Johnston. 2006. "*Corporate FTA (Futureoriented Technology Analysis).*" Anchor paper for the international FTA seminar, Seville.

Daheim, C., and G. Uerz. 2008. "Corporate Foresight in Europe: From Trend Based Logics to Open Foresight." *Technology Analysis & Strategic Management* 20 (3): 321–336.

Fikirkoca, A., and O. Saritas. 2012. "Foresight for Science Parks: The Case of Ankara University." *Technology Analysis & Strategic Management* 24 (10): 1071–1085.

Gavigan, J., and F. Scapolo. 2001. " Foresight and the Long-Term View for Regional Development." *The ITPS Report* 56: 19–29.

Gracht, H. A., C. R. Vennemann, and I.-L. Darcow. 2010. "Corporate Foresight and Innovation Management: A Portfolio-Approach in Evaluating Organizational Development." *Futures* 42: 380–393.

Grim, T. 2009. "Foresight Maturity Model (FMM): Achieving Best Practices in the Foresight Field." *Journal of Futures Studies* 13: 69–80.

Harmaakorpi, V., and T. Uotila. 2006. "Building Regional Visionary Capability. Futures Research in Resource-Based Regional Development." *Technological Forecasting and Social Change* 73 (7): 778–792.

Heger, T., and M. Boman. 2015. "Networked Foresight—The Case of EIT ICT Labs." *Technological Forecasting and Social Change* 101, 147–164.

Hervas-Oliver, Jose-Luis, and Jose Albors-Garrigos. 2008. "The Role of the Firm's Internal and Relational Capabilities in Clusters: When Distance and Embeddedness are not Enough to Explain Innovation." *Journal of Economic Geography* 9 (2): 263–283.

Hideg, E., E. Nováky, and P. Alács. 2014. "Interactive Foresight on the Hungarian SMEs." *Foresight (los Angeles, Calif)* 16 (4): 344–359.

Holmes, H., and M. Ferrill. 2005. "The Application of Operation and Technology Roadmapping to aid Singaporean SMEs Identify and Select Emerging Technologies." *Technological Forecasting and Social Change* 72 (3): 349–357.

Horton, A. 1999. "A Simple Guide to Successful Foresight." *Foresight (los Angeles, Calif)* 1: 5–9.

Jannek, K., and K. Burmeister. 2007. "Corporate Foresight in Small and Medium-sized Enterprises." EFMN Foresight Brief №101, Z-punkt, 2007.

Jun, S.-P., J. H. Seo, and J.-K. Son. 2013. "A Study of the SME Technology Roadmapping Program to Strengthen R&D Planning Capability of Korean SMEs." *Technological Forecasting and Social Change* 80 (5): 1002–1014.

Karaev, A., S. L. Koh, and L. T. Szamosi. 2007. "The Cluster Approach and SME Competitiveness: A Review." *Journal of Manufacturing Technology Management* 18 (7): 818–835.

Karasev, O., and K. Vishnevskiy. 2013. "A Toolkit for Integrated Roadmaps: Employing Nanotechnologies in Water and Wastewater Treatment." *Science, Technology and Innovation Policy for the Future*: 137–159.

Karasev, O., and K. Vishnevsky. 2010. "Identifying the Future of new Materials with the use of Foresight Methods." *Foresight-Russia* 4 (2): 58–67.

Keenan, M., S. Marvin, and C. Winters. 2002. *Mobilising the Regional Foresight Potential for Enlarged European Union.* Brussels: United Kingdom Country Report.

Keller, J., C. Markmann, and A. Heiko. 2015. "Foresight Support Systems to Facilitate Regional Innovations: A Conceptualization Case for a German Logistics Cluster." *Technological Forecasting and Social Change* 97: 15–28.

Kindras, A., D. Meissner, K. Vishnevskiy, and M. Cervantes. 2014. Regional Foresight for Bridging National Science, Technology and Innovation with Company Innovation: Experiences from Russia. Higher School of Economics Research Paper No. WP BRP, 29.

Kononiuk, A., and A. Sacio-Szymańska. 2015. "Assessing the Maturity Level of Foresight in Polish Companies—a Regional Perspective." *European Journal of Futures Research.* doi:10.1007/s40309-015-0082-9.

Kutsenko, E. 2015. "Pilot Innovative Territorial Clusters in Russia: A Sustainable Development Model." *Foresight-Russia* 9 (1): 32–55.

Lee, S., G. Park, B. Yoon, and J. Park. 2010. "Open Innovation in SMEs-An Intermediated Network Model." *Research policy* 39(2): 290-300.

Lichtenthaler, E., et al. 2004. "Organisation of the Early Phases of the Radical Innovation Process." *International Journal of Technology Intelligence and Planning* 1: 100–114.

Markmann, C., H. von der Gracht, J. Keller, and R. Kroehl. 2012, April. "*Collaborative Foresight as a Means to Face Future Risks—an Innovative Platform Conception.*" Proceedings of the 9th international ISCRAM conference, 1–5, The 10th Science and Technology Foresight Survey (NISTEP, 2015). NISTEP, 2015.

Nováky, E. 2010. "Alternative Scenarios for Hungary for the Year 2025." *Society and Economy* 32 (1): 47–64.

OECD. 2007. *Competitive Regional Clusters: National Policy Approaches.* Paris: OECD.

Oxford Research. 2008. *Cluster Policy in Europe. A Brief Summary of Cluster Policies in 31 European Countries, Kristiansand.* Norway: Oxford Research AS.

Paliokaite, A. 2010. *Networking as a Route for Corporate Foresight in SMEs,* IET Working Papers Series, WPS010.

Popper Villarroel, R., and R. Popper. 2008. "Foresight Methodology." In *The Handbook of Technology Foresight: Concepts and Practice,* edited by L. Georghiou, J. Cassingena, M. Keenan, and I. Miles, 1st ed., 44–88. Cheltenham: Edward Elgar.

Porter, M. E. 1998. "Clusters and the new Economics of Competition." *Harvard Bussiness Review* 76 (6): 77–90.

Poteralska, B., and A. Sacio-Szymańska. 2014. "Evaluation of Technology Foresight Projects." *European Journal of Futures Research.* doi:10.1007/s40309-013-0026-1.

Pozdnakova, A. 2008. "Knowledge Management and Foresight Elements for Small and medium-Sized Enterprises." *Computer Science* 34: 126–137.

Pro Inno Europe. 2012. *Key Messages and Practical Recommendations From the TACTICS Project.* Brussels: European Commission.

Rohrbeck, R. 2010. *Corporate Foresight: Towards a Maturity Model for the Future Orientation of a Firm.* Heidelberg: Physica-Verlag, Springer.

Rohrbeck, R., and H. G. Gemünden. 2011. "Corporate Foresight: Its Three Roles in Enhancing the Innovation Capacity of a Firm." *Technological Forecasting and Social Change* 78 (2): 231–243.

Rohrbeck, R., S. Mahdjour, S. Knab, and T. Frese. 2009. *Benchmarking Report: Strategic Foresight in Multinational Companies*. European Corporate Foresight Group.

Rohrbeck, R., and J. O. Schwarz. 2013. "The Value Contribution of Strategic Foresight: Insights From an Empirical Study of Large European Companies." *Technological Forecasting and Social Change* 80 (8): 1593–1606.

Salmenkaita, J. P., and A. Salo. 2004. "Emergent Foresight Processes: Industrial Activities in Wireless Communications." *Technological Forecasting and Social Change* 71 (9): 897–912.

Saritas, O. 2013. "Systemic Foresight Methodology." In *Science, Technology and Innovation Policy for the Future*, 83–117. Berlin: Springer.

Sarpong, D., M. Maclean, and E. Alexander. 2013. "Organizing Strategic Foresight: A Contextual Practice of 'way Finding'." *Futures* 53: 33–41.

Sato, Y. 2015. *Development of Small and Medium Enterprises in the ASEAN Economies*. Beyond 2015.

Savioz, P., and M. Blum. 2002. "Strategic Forecast Tool for SMEs: How the Opportunity Landscape Interacts with Business Strategy to Anticipate Technological Trends." *Technovation* 22 (2): 91–100.

Slaughter, R. A. 1999. "Futures Studies as an Intellectual and Applied Discipline." *American Behavioral Scientist* 42 (3): 372–385.

Sölvell, Ö, and M. Williams. 2013. *Building the Cluster Commons. An Evaluation of 12 Cluster Organizations in Sweden: 2005–2012*. Sweden: Ivory Tower Publishers.

Van de Vrande, V., J. Jong, and W. Vanhaverbeke. 2009. "Open Innovation in SMEs: Trends, Motives and Management Challenges." *Technovation* 29 (6): 423–437.

Vecchiato, R., and C. Roveda. 2010. "Strategic Foresight in Corporate Organizations: Handling the Effect and Response Uncertainty of Technology and Social Drivers of Change." *Technological Forecasting and Social Change* 77 (9): 1527–1539.

Vishnevskiy, K., and O. Egorova. 2015. "*Strategic Foresight for SMEs: Choice of the Most Relevant Methods*." Conference proceedings. The XXVI ISPIM conference – shaping the frontiers of innovation management, Budapest, Hungary, 2015.

Vishnevskiy, K., and O. Karasev. 2015. "Corporate Foresight in Russia: Challenges and Opportunities." In *Deploying Foresight for Policy and Strategy Makers - Turning Grand Challenges in Grand Opportunities Through Science*, edited by D. Meissner, L. Gokhberg, and A. Sokolov. Technology and Innovation, Springer.

Vishnevskiy, K., O. Karasev, and D. Meissner. 2015b. "Integrated Roadmaps and Corporate Foresight as Tools of Innovation Management: The Case of Russian Companies." *Technological Forecasting and Social Change* 90: 433–443.

Vishnevskiy, K., D. Meissner, and O. Karasev. 2015a. "Strategic Foresight: State-of-the-art and Prospects for Russian Corporations." *Foresight (los Angeles, Calif)* 17 (5): 460–474.

Overcoming open innovation challenges: a contribution from foresight and foresight networks

Jonathan Calof, Dirk Meissner and Anastasia Razheva

ABSTRACT

The article identifies how the combination of concepts drawn from foresight and foresight networks can be used to help open innovation. We found that foresight can support open innovation by providing analysis that looks at key open innovation questions such as those around technology selection, identifying future customer needs and scanning for disruptions. Foresight can also help open innovation address some of the challenges that have been identified in the open innovation literature as barriers to effective open innovation. Foresight has experience around obtaining access to appropriate external experts and their knowledge; making sense of the mass of information that can emerge through a more open process, both areas that the open innovation literature has identified as being challenges to effective open innovation. Finally, a concept explored in this paper, foresight networks offer's open innovation new ideas in innovative collaboration forms and how they can be pivotal in innovation and in assisting open innovation.

1. Introduction

Open innovation (OI) has emerged as a hot topic in academia and industry accompanied by the demand for the development of new approaches to take advantage of the need for integrated information networks both inside and outside the firm, which is at the heart of OI and also at the centre of foresight. Foresight and OI recognise that the knowledge required to make key decisions resides inside and outside the organisation.

OI research has identified several problems that impact organisations' ability to move forward with successful implementation for which as we will explain in this paper, foresight might have advice and solutions to offer OI:

1) attitudinal barriers within organisations related to accepting and seeking out external partners and networks (Salter, Criscuolo, and Ter Wal 2014);
2) the attraction of appropriate external experts to develop networks and gain contributions (Sieg, Wallin, and Von Krogh 2010; Parida, Oghazi, and Ericson 2014);
3) the integration and processing of overwhelming amounts of information (West and Gallagher 2006); and

4) the formulation of effective responses to intellectual-property challenges resulting from co-developments and situations where intellectual property needs to be opened up and shared in order to gain assistance from networks (Sieg, Wallin, and Von Krogh 2010; Parida, Oghazi, and Ericson 2014; Salter, Criscuolo, and Ter Wal 2014).

While the OI literature has developed various approaches to addressing these challenges, we take the next step of examining how the foresight field can effectively work with OI to both support it and to help them address the challenges identified in the literature. Much like OI, foresight focuses on the integration of knowledge from both inside and outside the organisation to assist with making innovation-related decisions (Cassingena Harper 2013). Its methodologies also address many of the same challenges described in the OI literature (Rohrbeck and Gemünden 2011). The richness of the foresight field, based on its state of development in terms of academic research, corporate and government use and based on its similarity in terms of external orientation, decision making and use of external experts, may provide OI with insights that can make OI more effective. Thus, this article explores how foresight and the more recently developed concept of foresight networks (FNWs) can both directly support OI and also provide advice, lessons learned and research results to address OI implementation challenges.

2. Setting the context: foresight and OI

Organisations are facing increasing pressure to adapt to the complexity and uncertainty of the business environment; these phenomena include the rise of new technologies that have caused some firms to lose their competitive advantage. Therefore, in order to survive and prosper, many organisations have had to use new innovation management instruments, including bottom-up, vision-driven approaches; these instruments have come to dominate these organisations' strategic innovation orientation (Schwarz 2008; Campanella et al. 2016; Trequattrini et al. 2016; Khripunova et al. 2014). Moreover, many firms have realised that innovation activities are more likely to be successful when they are conducted in cooperation with partners. This has led to the widespread use of OI that involves 'partnering' in all activities and stages of the innovation process (e.g. starting with idea collection and screening; actual development and production introduction; and the exploration of multiple exploitation paths).

In this regard it is important to acknowledge that company innovation activities have become more complex in all respects because more information has to be collected from both internal and external information networks and then has to be processed. This can become especially challenging during the early stages of the innovation process when the OI results associated with many ideas, exploitation paths and opportunities need to be prioritised within the context of various uncertainties and resource constraints (Meissner and Kotsemir 2016). Companies are increasingly engaging in collaborative innovation-development activities that frequently require partners with special competencies. Additionally, in this regard many companies are monitoring the global science, technology and innovation (STI) landscape in search of potential opportunities (Gokhberg and Meissner 2013; Cervantes and Meissner 2014; Choi and Choi 2015). The latter is commonly characterised as foresight activities conducted at different levels with the aim to remaining informed about potential STI developments and partners.

Among other strategic-planning tools and instruments, companies do conduct their own foresight exercises or similar activities to depict potential future developments and identify risks and opportunities that are related to their scope of activities (Del Giudice, Campanella, and Dezi 2016; Solima, Della Peruta, and Del Giudice 2016). This is referred to as 'corporate foresight.' Corporate foresight is more in depth and targeted than general (aggregate) foresight activities, which are much broader in scope and are conducted at an aggregate (e.g. national) level. Nevertheless, aggregate foresight offers companies multiple advantages, among which the validation of their own information and the networking element have proven to be especially valuable complementary resources. These values provide incentives for companies to participate in aggregate foresight, which in turn can lead to an increase in participants and in the scope and complexity of these types of activities. In part based on this understanding, foresight professionals have begun to organise activities in FNW, bringing together methodological and subject competencies and involving a broader range of stakeholders. Frequently these FNWs are spread over a country and are not necessarily concentrated in one region. Thus, foresight organised in respective networks involves more stakeholders at the national and regional levels, which is why FNW-building now appears to be one of the commonly discussed themes among managers and researchers alike.

The paper employs review of existing literature on Foresight, OI and Foresight Networks, analyzes and synthesises them and develops new conceptual thoughts. The literature review takes account of major publications in indexed journals. The literature review and conceptual work is complemented by authors' long standing experience in OI projects and foresight as well as FNW conceptualisation and implementation. For reasons of confidentiality authors can't disclose details of their involvement in these practical projects.

3. Evolution of foresight and foresight networks and the OI paradigm

3.1. Foresight

While open innovation traces its beginning to 2003 and Chesbrough's seminal work (Chesbrough 2003a), references to foresight go back to the early 1950s. Early examples of government-led foresight initiatives were identified in the 1950s (Bradfield et al. 2005). While Johnston (2008) identifies corporate-foresight initiatives in the 1960s including Dutch Royal Shell's year 2000 study initiated in 1967 and General Electric's year 2000 study done in 1971 formal definitions of foresight were articulated in the early 1990s. For example, Martin (1995) offered this definition of foresight in 1995

> ... the process involved in systematically attempting to look into the longer-term future of science, technology, the economy, environment and society with the aim of identifying the emerging generic technologies and the underpinning areas of strategic research likely to yield the greatest economic and social benefits (140).

Initially (during the middle of the twentieth century), foresight activities were primarily conducted by a small number of participants, largely from the interior of the organisation, using a limited number of approaches (e.g. a roadmap or a scenario). Over time these evolved to include a larger number of participants, greater use of external networks and a wider mix of analytical methodologies. Early foresight activities were designed for information and awareness-creation purposes and in some cases STI priority setting and evolved to become an integral part of public strategic STI planning (Meissner 2012, 2013a, 2013b) and corporate strategy. Georghiou (2001) describes this in terms of the evolution of foresight over three generations:

- First generation foresight: technology forecast;
- Second generation foresight: technology and markets; and
- Third generation foresight: technology, markets and the social dimension.

The evolution of foresight according to these authors from forecasting of technology development towards the integration of the market and social dimensions meant that foresight could be used more broadly in organisations but also imposed the requirement to include more diverse competencies and knowledge holders (experts) within foresight to address these new areas. The concept involves examining attempts to identify different generations, which entails the expansion of the foresight mission from its original, technology basis to the new configuration of FNW within different dimensions including the technology, market, social and political areas, as is illustrated in Figure 1 and as is reflected in the FNW domination.

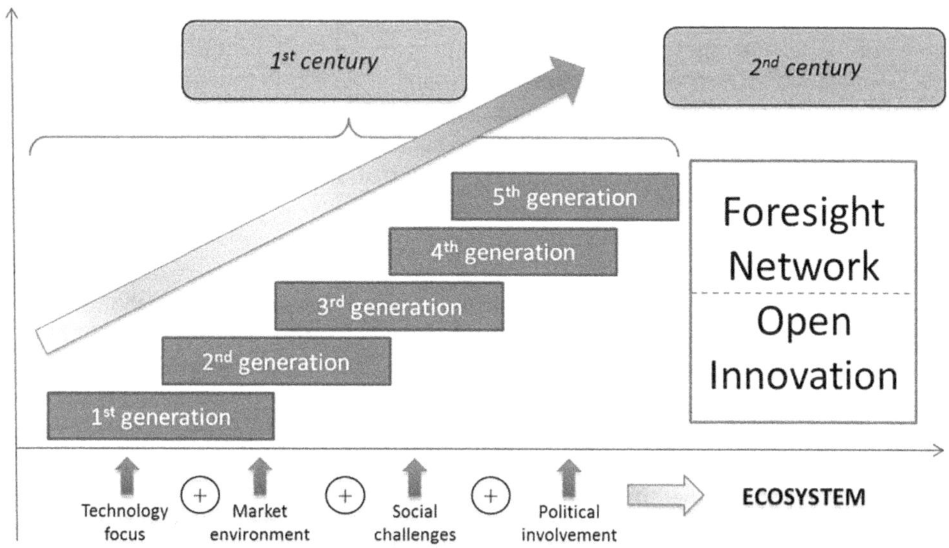

Figure 1. Development of the foresight system.
Source: National Research University Higher School of Economics based on Georghiou (2001).

With these developments in foresight, both in terms of breadth of application and depth of new approaches (methodologies) the definition of foresight also evolved to encompass the broadening use of foresight. Cassingena Harper (2013, 6), in referencing numerous researchers and authors in the foresight field, puts forwards a lengthy definition of foresight which in summary concludes that foresight is '… an approach for collectively exploring, anticipating, and shaping the future …'. Further she stresses several features of foresight, namely that it is '… systematic, participatory, future intelligence gathering and medium-to-long-term vision-building process aimed at present day decisions and mobilising joint actions'.

This definition incorporates the process/methodology of foresight, the important role played by stakeholders and external participants, and the ways in which foresight is used in supporting decision making (Meissner 2015). The objective of foresight has also broadened to include uncertainty reduction, the identification of opportunities and threats, the building of a consensus/common vision, the identification of priorities for research, potential changes in technologies, technology selection, market changes and so forth (Cassingena Harper 2013).

FNW as a concept was initiated only in the most recent five years. Van der Duin, Heger, and Schlesinger (2014) bring this concept up in the context of the networked foresight. They note that there has been an increase in collaboration over the past two decades which supports the concept of FNW's. FNWs imply both formal and informal communities formed by a group of individuals or institutions (e.g. universities, businesses, government bodies, customers) on the basis of cooperation for sharing experiences, promoting ideas, developing new technologies and/or creating new products aimed at joint visions in a community which is also labelled 'ecosystem' (Carayannis and Meissner 2016). Here FNW's can signify the beginning of a new age of domination of business development within an ecosystem. It shows that FNWs have the unique ability to stimulate innovation by identifying the potential for increasing, leveraging and maintaining the competitiveness of national economies. In a commercial industrial sense, such networks deliver unique value by making the chain of technology creation more transparent and by providing all of the necessary connections with the participants in this process.

One specific feature of FNWs is their strong orientation to the production not only of reports and support networks but also of technology or product roadmaps which are basic tools for the different stakeholders and their respective applications (Robertson, Swan, and Newell 1996). Indeed, a strong awareness of the technologies that have the potential to form the business environment in which one will be operating in the future is extremely important, especially in this era of global uncertainty. In this respect, it is obvious that FNWs are characterised by a strong commitment to the quality of work done in these networks; this makes them a valuable source of information and knowledge for validation and for the refinement of innovation-related strategies.

3.2. Open innovation

Jelinek et al. (2012) note that the number one issue framing R&D decisions and the ways in which those decisions will influence the R&D labs of the future is 'OI'. Moreover, 'Similar sentiments have been written in the management science field' (Hossain et al. 2016). Beth Comstock, the Chief Marketing Officer for General Electric, summed up the imperative for OI when she explained that 'These problems are too big for us to solve alone. We need to collaborate like we never have before' (Nagji and Walters 2012).

The core of OI is the idea that the traditional closed model of innovation, which relied on in-house idea generation, conversion using internal personnel and company-controlled diffusion to market, could and would benefit from opening up the entire innovation value chain to external partners (Carayannis and Meissner 2016). Increasing R&D costs, increasing mobility of knowledge workers, expanding availability of venture capital funds, shortening product life cycles and increasing competition combined to seriously undermine the viability of a fully closed innovation model (Chesbrough 2003a).

Much like foresight, OI relies on external and internal input, as it is an externally oriented discipline that opens the organisation's decision-making systems to external collaborators. The OI literature has undergone extensive development, especially since 2009 (Hossain et al. 2016; West and Bogers 2014). Extensive research has looked at the barriers to implementing OI finding and, in particular, at the importance of and at approaches to dealing with attitudinal barriers, attracting and managing external contributors, managing intellectual property, and developing an appropriate organisational capability.

Attitudes: Consistent with the initial writings of Chesbrough (2003a), many of the barriers lie in the attitude shift required to move from fully controlling innovation within the organisation to integrating external partners into the process, which is generally referred to as 'not invented here'. Other attitudes highlighted in the OI literature include risk aversion, the level of willingness to trust external partners, the level of internal commitment from scientists and managers to accept innovative ideas from the outside, resistance to change and rigid organisational cultures (Sieg, Wallin, and Von Krogh 2010; Parida, Oghazi, and Ericson 2014).

External contributors: Two principal problems have been highlighted in the literature with regard to external contributors:

1. Ensuring that the right experts (i.e. those with the requisite knowledge) consent to be involved with the organisations (Sieg, Wallin, and Von Krogh 2010; Parida, Oghazi, and Ericson 2014).
2. Ensuring that, once they have agreed to become involved, these experts actually share their knowledge (West and Gallagher 2006).

After being attracted and convinced to share their information, companies are challenged with the opportunistic behaviour of external participants and with the need to manage cultural differences, foreign laws, strategic interests and other matters (Parida, Oghazi, and Ericson 2014).

Foresight has had to deal with these two issues and has developed practices around them. Therefore foresight can help OI address these two critical areas of OI challenges/barriers. This will be discussed in greater detail in the next section of this paper; it provides an opportunity for foresight to contribute to the OI field.

Management of intellectual property: Two major concerns that dominate the discussion are set out as follows:

1) OI requires that the organisation share details of existing IP and intended IP to enable partners to effectively contribute to the innovation process. As such, many companies express concerns about the dangers of revealing IP and specifically about the risk that it could be disseminated to a broader external environment including to competitors (West and Gallagher 2006; Sieg, Wallin, and Von Krogh 2010; Parida, Oghazi, and Ericson 2014; Salter, Criscuolo, and Ter Wal 2014).
2) Concerns about how to manage the IP that is created through OI. OI results in the development of new intellectual property and this raises the issue of how this new IP is effectively shared (Salter, Criscuolo, and Ter Wal 2014).

Organisational barriers: Having the absorptive capacity and the ability to recognise, integrate and use external contributions to the organisation's innovation process has also been a dominant theme within the OI literature and the foresight literature. While this dimension includes some of the attitudes mentioned above, it also involves reward mechanisms that do not favour internal development over external development, flexible and adaptable structures, access to scientists and others who have been trained to develop technology with partners, and appropriate information systems that are capable of integrating information from external sources without producing information overload and on a timely basis (West and Gallagher 2006; Parida, Oghazi, and Ericson 2014; Salter, Criscuolo, and Ter Wal 2014; Bondarenko 2015; Havas 2016).

These two latter areas are also addressed by foresight. Once again, studies on foresight's critical success factors have highlighted the organisational competencies required and the ways in which they can be developed. Another aspect of foresight development that may help to address some of the problems associated with OI deals with managing the substantial information inflows through OI and with identifying and selecting the most appropriate ideas (e.g. Sieg, Wallin, and Von Krogh 2010).

4. Foresight's potential contribution to open innovation

In reviewing the literature and developments within OI, foresight and FNW, it is evident that all three have a strong orientation towards collaboration. All of these approaches require appropriate absorptive capacity, attitudes that value and in fact seek out external expertise and knowledge, and an ability to attract appropriate external experts and provide them with the opportunity to share their information with the organisation. Foresight conceptually is focused on understanding the external environment (Calof and Smith 2010) using as primary methodologies approaches that involve integrating knowledge from individuals and entities outside the organisation. Similarly, OI focuses on integrating information from individuals and entities located outside the organisation to assist primarily with making technology and product decisions. While there are certainly important differences between them, there are various similarities and ways in which each field can assist and support the other. It is interesting to note that in the literature review done for this paper, we identified articles and case studies of dozens of companies that had both foresight units and OI initiatives (here are a few of them, IBM, P&G, Siemens, Volvo, Ericsson, Lego and many more). Uerz and Rudzinski (2012, 298) in their study on foresight and OI at Volkswagen wrote about how a company can benefit by having both a foresight and OI programme working together:

> The combination of two established methods out of the innovation management repertoire – the Scenario Thinking approach and Information Markets – is able to generate strategic insights/value and to establish a long-term perspective on strategy, embedded within an iterative learning cycle on strategy and innovation. Thus, the chosen approach of combining OI and foresight is able to serve both, the long term strategic requirements of a large organization and the more short-term oriented need to foster innovation at the same time.

Thus, in the case of Volkswagen it appears that foresight and OI offer a complementary approach that when used together fosters innovation.

How can foresight help innovation in general and OI in particular? Several researchers have provided insights into foresights role in enabling innovation (Paliokaitė and Pačėsa 2015). Öner and Göl Beşer (2011) wrote that 'Corporate foresight methods and activities are becoming crucial for innovation processes' (51). Ruff (2006) notes that, through foresight techniques such as Triz and Scenarios, several different focuses of corporate foresight could be identified and combined in order to contribute to effective innovation: 'long-term market developments, future customer needs, prospective evaluation of innovations, business processes and organisational change, and scanning and monitoring.' (278). Sarpong and Maclain (2011) observed companies using scenario's (a popular foresight technique) for the identification of opportunities for innovation reporting on 'a direct link between scenario planning/thinking and innovation' (1154). Von der Gracht and Stillings (2013) in their study of innovation-focused scenario's noted based on their research and past research that corporate foresight can increase the innovative capacity of companies. In their study they used three foresight approaches within a scenario process – Delphi expert survey, participatory workshops, and an open network approach in order to increase innovative capacity at a large global manufacturer (Miles, Saritas, and Sokolov 2016). Rohrbeck and Gemünden (2011) link foresight to many aspects of innovation and identify three specific roles for foresight in innovation:

- the strategist's role in directing innovation activities;
- the initiator's role in triggering innovation initiatives; and
- the opponent's role in challenging innovators.

Cassingena Harper (2013) looks at foresight's role in innovation more at the government level than in terms of corporate foresight and articulates similar conclusions, including the idea that foresight is prominently applied in the formulation of innovation policy and strategy. It has been recognised that foresight methodologies and foresight itself can be used to examine many of the same issues that are mentioned in the OI literature, such as understanding the impact of social change, determining market directions in the future and choosing appropriate technologies (Rohrbeck 2010). These and other authors point to foresight as a tool to inform the innovation process. Foresight encompasses analytical methodologies that can be used to assist in the process of making various kinds of technology, product and R&D decisions. OI lists as challenges understanding customer needs and having appropriate market intelligence and technical knowledge.

As can be seen from the discussion set out above, these are a reason which foresight is well equipped to provide guidance and for which foresight has analytical methods for addressing these needs. In this way, foresight can support OI by providing approaches and methodologies that helps OI through several stages of the innovation lifecycle. For example in the conversion phase a foresight project could use Delphi to determine what the best ideas are, scenarios could also be used (Miles, Saritas, and Sokolov 2016). In shaping idea generation, long term foresight project that look at evolving markets, technology disruption and societal trends could be done to help those generating ideas do so within the context of the results of the foresight studies.

In exploring other potential contributions that foresight can provide to OI, it is worth noting that OI's key problems and challenges (discussed in the previous section), namely attitudes towards external information and involvement from appropriate external participants, are issues that foresight has had to address by virtue of its externally oriented collaborative methodologies. Idea management – that is, how to select ideas and deal with massive information inflows from integrating external contributors (West and Gallagher 2006) – are also areas where foresight research and methodologies have found solutions. But perhaps a larger contribution between foresight and OI lies in the emerging area of FNWs and in the role of developing, creating, and effectively using knowledge and expert networks. This is explored in more detail in the next section with a specific focus on the kinds of collaboration and collaboration vehicles that can arise through this approach.

5. Collaboration between FNWs and OI

The previous section looked at the challenges identified in implementing OI and at the role that foresight can play in helping to overcome some of these challenges. This section looks at how FNW can be used to help implement OI. The evolving concept of FNW, when combined with other concepts in OI, offers approaches for developing more effective networks – something that is critical to both foresight and innovation. Overall, much like FNW, the OI paradigm similarly looks at absorbing industrial, academic and government knowledge to assist with the strategic-planning process at the business or government level. Players within the ecosystem seek to reduce their transaction costs which allows for easier access to knowledge (Carayannis and Meissner 2016). This in turn facilitates innovation and emerging technologies, thus cushioning the risks and the negative consequences that can accompany uncertainty. Collaboration between these different levels (industrial, government, academic and so forth) can be looked at in terms of the relationship between the models of OI and foresight from several different perspectives or approaches.

The *industrial approach* is oriented towards forming the FNW within or between industries with the goal of forming joint visions of their development. The strength of collaboration lies in the identification of opportunities for technological change as well as in the identification and development of breakthrough technologies (Johnston 2007). However, shortening technology life cycles and high level of diversity of companies within an industry might make this type of collaboration unsustainable. Structurally and organisationally, collaboration from the industrial approach is usually in the form of development agencies, joint development partnerships, joint technology initiatives and regional clusters. Joint foresight efforts might bring significant results when developing industrial or technology roadmaps under an industrial approach by combining the knowledge of the organisations within the industrial network.

The *knowledge-based approach* suggests that the sharing of knowledge and experience and the identification of the most promising areas of science and technology within the business, industrial and academic communities are essential for maximising the benefits that can be obtained from collaboration (Miles 2011). This approach is sometimes criticised from the standpoint of the gaps between consumer needs, technological capacities and scientific research. In addition, it requires appropriate regulation for the protection of intellectual property. Organisationally and structurally, this kind of collaboration has been seen in the form of R&D centres, the creation of the departments of enterprise at universities, and other forms of educational and business cooperation (Laursen and Salter, 2004). Joint foresight in the form of knowledge-management roadmaps is a useful instrument for understanding the full picture related to the use of knowledge and its dissemination within the knowledge-based approach (Calof and Smith 2012).

The *strategy-oriented approach* deals with collaboration from the perspective of the strategic-planning process within the firm. Collaboration is widely implemented for the identification of prospective market niches and product lines, the creation of communication platforms, and the analysis of the competitive environment. It does, however, bring certain risks to the organisation in terms of opportunistic behaviour, transaction costs (related to coordination and searches for a partner), ineffective sharing of risks, the theft of intellectual property, related external effects and so forth (Gassmann, Enkel, and Chesbrough 2010). Organisational and structural forms of collaboration under the strategy-oriented approach include spin-offs, strategic alliances, joint ventures and customer integration. There are numerous examples of the successful creation and implementation of business and product roadmaps as a result of these kinds of corporate foresight efforts.

The *technology approach* implies that firms are jointly collaborating with others for the purposes of identifying breakthrough technologies, optimising their technology value chains, setting their technology priorities and so forth (Yang and Wu 2010). Organisational and structural forms of collaboration under the technology approach include joint technology initiatives, joint development partnerships, development agencies and technology platforms.

The *transaction-costs approach* considers the key benefits that can be derived from collaboration in terms of decreasing the level of uncertainty and sharing risks (Kortelainen, Kutvonen, and Torkkeli 2012). This has been identified as a benefit of both foresight and OI (Chesbrough 2003b). In due course, the lower level of uncertainty allows for better access to new knowledge, for the implementation of new practices, for the reduction of the distance between the company and its customers, and so on. Taken together, all of these fruits contribute to the greater effectiveness of priority setting and decision making. However, the inherent risk of the theft of intellectual property remains prominent in the mind of each participant in the innovation-creation process, along with bargaining costs, opportunistic costs and the potential for conflicts of interest (Langlois 1992).

Summarising the material set out above, Table 1 provides forms for innovative collaboration with respect to the proposed approaches. Each of the approaches discussed in this section is accompanied by different types of networks and different foresight approaches as well as different forms of collaboration.

One final note on the approaches and forms of collaboration, from a transaction costs perspective, all forms of collaboration can be effective to a certain degree in decreasing uncertainty and sharing risks.

Mentioned in an earlier section of the paper is the concern articulated in the OI literature regarding the potential loss of intellectual property (IP). Given the fact that FNW is focused on the establishment of different kinds of networks, IP would be a major issue for FNW as well. The FNW approach, which involves the creation of broader networks, might appear to be at odds with competitive advantage concerns that arise when there is a danger of loss of IP to competitors. However, in order to accelerate the process of innovation creation, it is reasonable and sometimes necessary to share ideas, even with competitors. Further, one could argue that, under an FNW approach, network participants who engage in opportunistic behaviour and who appropriate IP from the group without proper permission may lose the future benefits of membership participation. Thus, the results of FNWs might be spread among the maximum number of beneficiaries who can collaborate in order to achieve a significant return. Despite the concerns around IP protection, it is clear that, with the significant growth in OI and the successes that have been identified (even when working with competitors) IP issues may not be as significant as first thought in the literature.

Table 1. Effectiveness of the forms of innovative collaboration from the standpoints of different approaches.

Collaboration forms	Industrial approach	Knowledge-based approach	Strategy-oriented approach	Technology approach	Transaction-costs approach
Customer integration			X		X
Educational and business cooperation		X			X
Joint development partnerships	X		X	X	X
Joint technology initiatives	X			X	X
Joint ventures			X		X
R&D centres		X			X
Regional clusters	X				X
Strategic alliances			X		X
Technology platforms	X			X	X
Collaborative roadmapping	Knowledge-management roadmaps	Industrial roadmaps, technological roadmaps	Business roadmaps, product roadmaps	Technology roadmaps	Any type of roadmap

Source: National Research University Higher School of Economics.

6. The impact of OI, foresight and FNW linkages

This paper has reported on several studies that have discussed the relationship between innovation in general on the one hand and OI and foresight including their similarities and how they support each other on the other hand (e.g. Rohrbeck 2010). Interestingly, similar driving forces have been found in the literature for all three of these approaches. This is summarised in Table 2. It follows that FNW/foresight and OI are highly complementary. Both have elements of external orientation and share techniques and methodologies that, when taken together, can be of enormous value to innovation performance and firm performance as well as to the development of government policies related to innovation and technology.

One of the main characteristics is that the OI model, foresight and FNW involve a large number of players for cooperation at the industry, government and academic levels – players that are eventually acting as open innovators. Another common feature concerns the coordination process for applying innovation tools by open innovators that are dependent on a particular purpose, which thus might be considered as foresight services.

Numerous actors within ecosystems might follow the new innovation paradigm. To be clear, OI requires innovators to form a holistic view of the world (within the scientific, technology, market and social dimensions) that can be enhanced by FNWs that aim at providing structured information drawn from multiplied sources. In this sense, there are several features that are shared between an OI model and FNW: high R&D costs, a long time horizon, a high level of uncertainty, open-minded behaviour, and active participation in and contributions to networking.

The rationale for implementing OI and foresight concepts is based on assumptions about the main benefits that can be derived from the implementation of those concepts. The most significant impacts of foresight on an OI model and vice versa are shown in Figure 2.

By identifying the outer limits of the OI model and FNW, the key links between these two concepts are defined. Figure 3 shows that a company might use foresight for solving a wide range of challenges. In particular, it can allow a company to focus on its identified technology priorities, to open up new market opportunities, to reveal breakthrough technologies, to foster the development of prospective product lines, to enhance the creation of communication platforms, to establish an expert base, and thus to reduce its cost inefficiencies and to define its optimal value chains. At the same time, a company might benefit from the implementation of an OI model, in part based on results obtained from foresight. This can enable a company to share risks, to expand innovation output, and to enhance inter- and intra-country technological change. Furthermore, it can lead to a decrease in R&D costs, to a reduction in time to market, and to the sharing of facilities and competencies, as well as helping to tackle any potential environmental uncertainty. By using the model, a company has an enhanced ability to identify new business niches and models. Regardless, it allows for collaboration in some of the most efficient ways based on consolidated opinions and based on identified visions by choosing the most appropriate form of collaboration for each innovation

Table 2. Driving forces of foresight, OI and FNWs.

Factor	Foresight	OI	FNWs
Rising uncertainty and risks	**	**	**
Higher level of openness and disclosure	*	**	**
Increasing information flows	*	*	**
New paradigm of the planning process	**	**	*
Detecting emerging breakthrough technologies	**	**	**
Increasing demand needs	*	**	*
Major challenges – climate change, food and energy security, etc.	**	*	**
Cooperation between business, government and academia	*	**	**
Increasing role of knowledge and innovation	*	**	*
Tightening market competition, searching for competitive advantages	**	**	**

Note: *Medium influence, **High influence.
Source: National Research University Higher School of Economics.

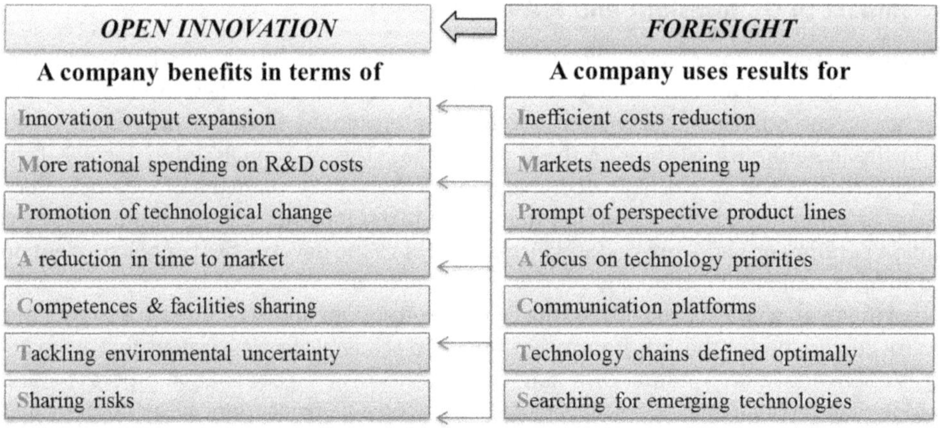

Figure 2. The impacts of foresight results on the OI model.
Source: National Research University Higher School of Economics.

management phase. Simultaneously, an open innovator that aims to use foresight in its strategic-planning process would benefit from creating common visions and consolidating opinions, from elaborating strategic guidance, and from analyzing competitors' concepts, customer needs, and the political and regulatory environment (Rohrbeck and Gemünden 2011). It could also allow for the identification of promising scientific and technological areas, and for undertaking the comparative analysis of domestic and foreign science and technology achievements.

As far as ecosystem applications are concerned, the level of impact on the innovation framework, the level of involvement of government representatives and the power of the state lobby are considered. In terms of business purposes, a stronger interest emerges to create joint ventures, strategic alliances and spin-offs, as well as to cooperate with consumers. Also for other actors that have an interest in industry, it might be reasonable to participate in regional clusters, in educational and business cooperation and in probably joint development partnerships as this can enhance inter-industrial collaboration. The analysis proves that all of the above-mentioned forms of innovative collaboration contribute to a certain extent to the creation and development of an FNW.

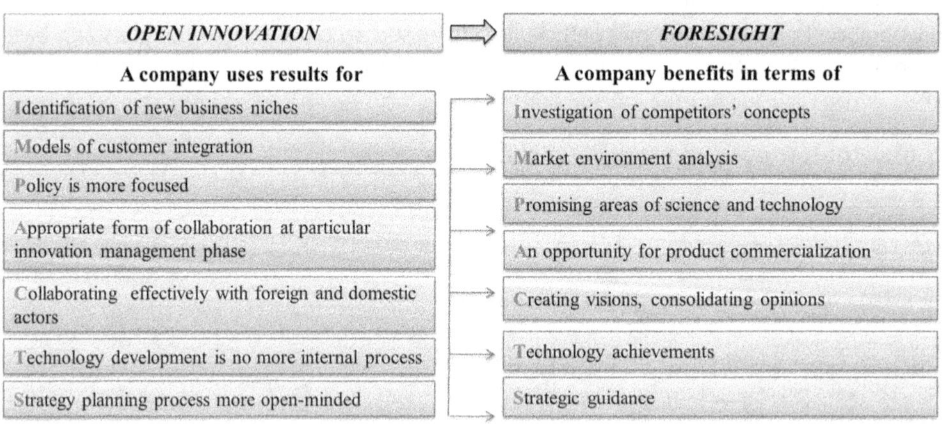

Figure 3. The impacts of the OI model results on foresight.
Source: National Research University Higher School of Economics.

7. Conclusions and discussion

This paper has looked at how foresight can help open innovation. It has addressed it from three broad perspectives. First in terms of how foresight can directly support open innovation. Secondly, in terms of how the lessons learned through generations of foresight can help open innovation address challenges and barriers facing OI. Third it has looked at how the concept of FNW can help OI's external oriented partnership based orientation. Through the literature review and the development of a collaboration model, this paper has identified several ways in which foresight can help OI. The following summarises what has been written in this paper in terms of how foresight can help OI.

Foresight analytical techniques and methods can address many of the questions posed in the OI literature such as identifying technology directions and understanding social and market trends. Techniques like scenarios, triz and roadmapping are well suited to these purposes.

Also Foresight has developed approaches for integrating opinions/knowledge from large numbers of external and internal experts. Approaches like Delphi could help OI both to integrate this kind of information and to enlist the support of the expert group in helping to make choices and decisions. Several of the organisational attributes associated with OI's absorptive capacity have been found in the foresight literature. Thus, organisational-design elements including training programmes and project design elements drawn from the foresight literature and from project experience could be of assistance to OI.

Attracting appropriate experts and encouraging them to provide their information are at the core of many foresight techniques. Accordingly, there is extensive foresight knowledge and experience associated with addressing the concerns in OI around managing the challenges in these areas.

1. Sharing knowledge, including advanced knowledge of IP and even joint IP development, is a part of the legacy of foresight, and is an element for consideration given the methodologies that are used. Accordingly, much like the previous point (attracting experts), there is considerable knowledge and experience within the foresight community on how these issues have been dealt with.

Clearly, foresight methods and approaches can help OI, especially in areas related to technology selection. As well, this paper has also shown how FNW can help OI through better collaboration. Various forms of collaboration were described and then were matched to the type of network being formed.

In looking at foresight tools, techniques and approaches as well as at FNW within an OI framework, it becomes possible to appreciate foresight's involvement throughout all of the stages of an OI initiative. A good starting point for an OI project could be a series of foresight workshops aimed at forming a framework, gathering information and setting priorities based on the analysis of risks, barriers, challenges and trends in the relevant area of science and technology. In addition to priority setting, the focal points for growth and windows of opportunity could be considered (Archibugia and Pietrobelli 2008). Figure 4 combines an illustration of diverse forms of OI models disseminated through Innovation-development phases on the one hand and foresight tasks that are appropriate to the various stages of the innovation process on the other hand. This model follows from the discussion in this paper and should be tested in future research.

In bringing together all the ideas brought up in this paper, how foresight can support OI through its methodologies, the FNW concept and collaborative platforms we offer to the reader the following view of how these elements could work together (Figure 4). This is a conceptual model that flows from the concepts advanced in this paper and will need to be further developed and tested in the future.

At the stage of generating and applying new ideas, open innovators can contribute by participating in technology platforms, R&D centres and crowd-sourcing platforms while placing a strong

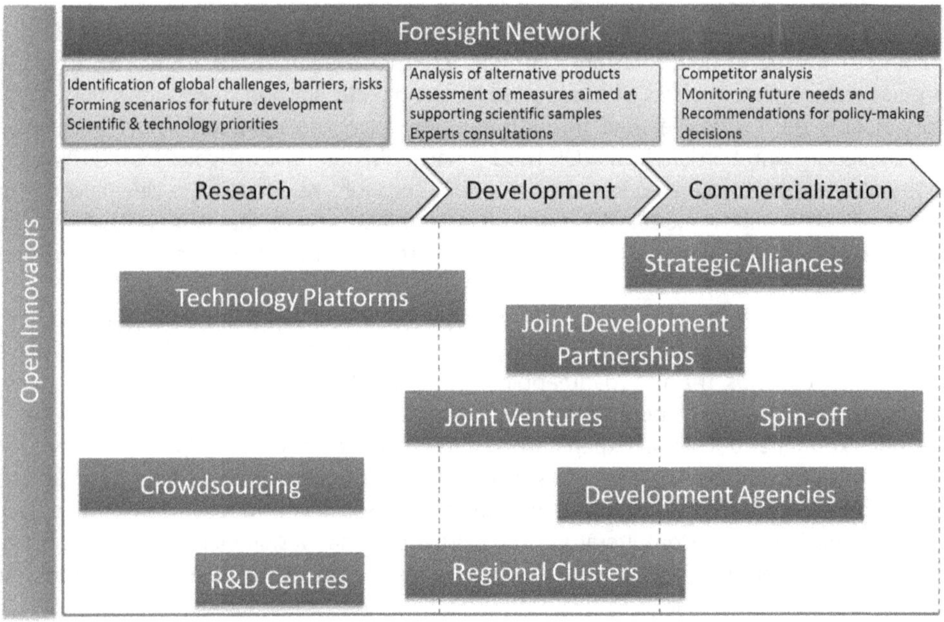

Figure 4. Accompaniment of the innovation-management phases by open innovators and FNWs.
Source: National Research University Higher School of Economics.

emphasis on global challenges, barriers and risks, as well as on the formulation of future goals. The next stage is supposed to be followed by collaboration within joint ventures and innovation clusters that are well suited to undertaking measures for obtaining scientific samples and searching for alternatives. Moving towards commercialisation, a company might cooperate and collaborate with interested parties in different ways by focusing on competitor analysis, market assessment, and recommendations for policy-makers. Therefore, it should be underlined that the primary role assigned to foresight appears in the stage of basic and applied research, while various forms of OI are mainly concentrated on the phases of development and commercialisation. Nevertheless, both OI collaboration and FNW should accompany the entire innovation lifecycle, though the respective roles played by OI and foresight might change while pursuing science and technology policies that are applied to different product groups.

The purpose of this paper was to identify ways in which foresight and FNWs could help address OI challenges. It has identified how these can be used at each stage of the innovation value chain and management process and how foresight and FNW research and experiences can be used to address the challenges to OI implementation that have been identified in the OI literature. Interestingly, the literature review revealed that the explored subject is highly relevant and that it has been investigated by many researchers in the innovation field and in related fields.

Disclosure statement

No potential conflict of interest was reported by the authors.

Funding

The article was prepared within the framework of the Basic Research Programme at the National Research University Higher School of Economics (HSE) and supported within the framework of the subsidy granted to the HSE by the Government of the Russian Federation for the implementation of the Global Competitiveness Programme.

ORCID

Jonathan Calof http://orcid.org/0000-0002-2146-1506
Dirk Meissner http://orcid.org/0000-0003-4573-0376

References

Archibugia, D., and C. Pietrobelli. 2008. "Knowledge Sharing in an Organization." *Technological Forecasting and Social Change* 75 (8): 1128–1156.

Bondarenko, N. 2015. "The Role of Companies in Human Capital Accumulation: Cross-Country Analysis." *Foresight-Russia* 9 (2): 22–37.

Bradfield, R., G. Wright, G. Burt, G. Cairns, and K. Van Der Heijden. 2005. "The Origins and Evolution of Scenario Techniques in Long Range Business Planning." *Futures* 37 (8): 795–812.

Calof, J., and J. Smith. 2010. "Critical Success Factors for Government-Led Foresight." *Science and Public Policy* 37 (1): 31–40.

Calof, J., and J. Smith. 2012. "Foresight Impacts from Around the World: A Special Issue." *Foresight* 14 (1): 5–14.

Campanella, F., M. Del Giudice, A. Thrassou, and D. Vrontis. 2016. "Ambidextrous Organizations in the Banking Sector: An Empirical Verification of Banks' Performance and Conceptual Development." *The International Journal of Human Resource Management*, 1–31.

Carayannis, E. G., and D. Meissner. 2016. "Glocal Targeted Open Innovation: Challenges, Opportunities and Implications for Theory, Policy and Practice." *The Journal of Technology Transfer* 2016: 1–17. doi:10.1007/s10961-016-9497-0.

Cassingena Harper, J. 2013. "Impact of Technology Foresight on Innovation Policy." Innovation-policy.org.uk. Accessed August 9, 2016. http://www.innovation-policy.org.uk/compendium/section/Default.aspx?topicid=31

Cervantes, M., and D. Meissner. 2014. "Commercialising Public Research Under the Open Innovation Model: New Trends." *Foresight and STI Governance* 8 (3): 70–81.

Chesbrough, H. 2003a. "The Era of Open Innovation." *MIT Sloan Management Review* 44 (3): 35–41.

Chesbrough, H. 2003b. *Open Innovation: The New Imperative for Creating and Profiting from Technology*. Boston, MA: Harvard Business School Press.

Choi, M., and H. Choi. 2015. "Foresight for Science and Technology Priority Setting in Korea." *Foresight and STI Governance* 9 (3): 54–67.

Del Giudice, M., F. Campanella, and L. Dezi. 2016. "The Bank of Things: An Empirical Investigation on the Profitability of the Financial Services of the Future." *Business Process Management Journal* 22 (2): 324–340.

Gassmann, O., E. Enkel, and H. Chesbrough. 2010. "The Future of Open Innovation." *R&D Management* 40 (3): 213–221.

Georghiou, L. 2001. "Third Generation Foresight –Integrating the Socio-Economic Dimension." Conference Contribution, NISTEP, Japan.

Gokhberg, L., and D. Meissner. 2013. "Innovation: Superpowered Invention." *Nature* 501: 313–314.

Havas, A. 2016. "Social and Business Innovations: Are Common Measurement Approaches Possible?" *Foresight and STI Governance* 10 (2): 58–80.

Hossain, M., K. Islam, M. Sayeed, and I. Kauranen. 2016. "A Comprehensive Review of Open Innovation Literature." *Journal of Science and Technology Policy Management* 7 (1): 2–25.

Jelinek, M., A. Bean, R. Antcliff, E. Whalen-Pedersen, and A. Cantwell. 2012. "21st-Century R&D: New Rules and Roles for the R&D "Lab" of the Environment." *Research-Technology Management* 55: 16–26.

Johnston, R. 2007. "Priority Setting for Future Critical and Key Industrial Technologies as Driving Forces for Economic Development and Competitiveness." UNIDO technology foresight summit, Budapest, Hungary, September 27–29.

Johnston, R. 2008. "Historical Review of the Development of Future-Oriented Technology Analysis." In *Future-Oriented Technology Analysis: Strategic Intelligence for an Innovative Economy*, edited by C. Cagnin, M. Keenan, R. Johnston, F. Scapolo and R. Barre, 1st ed., 17–23. Berlin: Springer-Verlag.

Khripunova, A., K. Vishnevskiy, O. Karasev, and D. Meissner. 2014. "Corporate Foresight for Corporate Functions: Impacts From Purchasing Functions." *Strategic Change* 23: 147–160.

Kortelainen, S., A. Kutvonen, and M. Torkkeli. 2012. "Open Innovation Versus Transaction Cost Economics." Proceedings of ISPIM conferences, Barcelona, June 17–20.

Langlois, R. N. 1992. "Transaction-Cost Economics in Real Time." *Industrial and Corporate Change* 1: 99–127.

Laursen, K., and A. Salter. 2004. "Searching High and Low: What Types of Firms Use Universities as a Source of Innovation?" *Research Policy* 33 (8): 1201–1215.

Martin, B. R. 1995. "Foresight in Science and Technology." *Technology Analysis & Strategic Management* 7 (2): 139–168.

Meissner, D. 2012. "Results and Impact of National Foresight-Studies." *Futures* 44 (10): 905–913.

Meissner, D. 2013a. "Results and Impact of National Foresight-Studies." In *Science, Technology and Innovation Policy for the Future – Potentials and Limits of Foresight Studies*, edited by D. Meissner, L. Gokhberg and A. Sokolov, 31–41. London: Springer.

Meissner, D. 2013b. "Instruments to Measure Foresight." In *Science, Technology and Innovation Policy for the Future – Potentials and Limits of Foresight Studies*, edited by D. Meissner, L. Gokhberg and A. Sokolov, 43–62. London: Springer.

Meissner, D. 2015. "Developing 'Green Thinking' Towards Sustainability." *International Journal of Social Ecology and Sustainable Development* 6 (3): iv–vii.

Meissner, D., and M. Kotsemir. 2016. "Conceptualizing the Innovation Process Towards the 'Active Innovation Paradigm'—Trends and Outlook." *Journal of Innovation and Entrepreneurship A Systems View Across Time and Space 2016* 5: 14. doi:10.1186/s13731-016-0042-z.

Miles, I. 2011. "From Knowledge-Intensive Services to Knowledge-Intensive Service Systems." *International Journal of Services Technology and Management* 16 (2): 141–159.

Miles, I., O. Saritas, and A. Sokolov. 2016. *Foresight for Science, Technology and Innovation*. Heidelberg: Springer International Publishing.

Nagji, B., and H. Walters. 2012. "Help Wanted 2.0: Engaging Others to Tackle Wicked Problems." *Rotman Magazine* Spring: 73–76.

Öner, A., and S. Göl Beşer. 2011. "Assessment of Corporate Foresight Project Results: Case of a Multinational Company in Turkey." *Foresight* 13 (2): 49–63.

Paliokaitė, A., and N. Pačėsa. 2015. "The Relationship Between Organisational Foresight and Organisational Ambidexterity." *Technological Forecasting and Social Change* 101: 165–181.

Parida, V., P. Oghazi, and Å Ericson. 2014. "Realization of Open Innovation: A Case Study in the Manufacturing Industry." *Journal of Promotion Management* 20 (3): 372–389.

Robertson, M., J. Swan, and S. Newell. 1996. "The Role of Networks in the Diffusion of Technological Innovation." *Journal of Management Studies* 33 (3): 333–359.

Rohrbeck, R. 2010. "Harnessing a Network of Experts for Competitive Advantage: Technology Scouting in the ICT Industry." *R&D Management* 40 (2): 169–180.

Rohrbeck, R., and H. G. Gemünden. 2011. "Corporate Foresight: Its Three Roles in Enhancing the Innovation Capacity of a Firm." *Technological Forecasting and Social Change* 78 (2): 231–243.

Salter, A., P. Criscuolo, and A. Ter Wal. 2014. "Coping with Open Innovation: Responding to the Challenges of External Engagement in R&D." *California Management Review* 56 (2): 77–94.

Sarpong, D., and M. Maclain. 2011. "Scenario Thinking: A Practice-Based Approach for the Identification of Opportunities for Innovation." *Futures,* 43: 1154–1163.

Schwarz, J. O. 2008. "Assessing the Future of Futures Studies in Management." *Futures* 40 (3): 237–246.

Sieg, J., M. Wallin, and G. Von Krogh. 2010. "Managerial Challenges in Open Innovation: A Study of Innovation Intermediation in the Chemical Industry." *R&D Management* 40 (3): 281–291.

Solima, L., M. R. Della Peruta, and M. Del Giudice. 2016. "Object-Generated Content and Knowledge Sharing: The Forthcoming Impact of the Internet of Things." *Journal of the Knowledge Economy* 7 (3): 738–752.

Trequattrini, R., M. Del Giudice, B. Cuozzo, and M. Palmaccio. 2016. "Does Sport Innovation Create Value? The Case of Professional Football Clubs." *Technology, Innovation and Education* 2 (1): 1–15.

Uerz, G., and C. Rudzinski. 2012. "Foresight & Open Innovation at Volkswagen: Creating Strategic Value by Combining the Scenario & Information Market Approach." In *Proceedings of ISPIM Conferences 23*, 1–14. Manchester: Society for Professional Innovation Management (ISPIM).

Van der Duin, P., T. Heger, and M. D. Schlesinger. 2014. "Towards Networked Foresight? Exploring the Use of Futures Research in Innovation Networks." *Futures* 59 (June): 62–78.

Von der Gracht, H. A., and C. Stillings. 2013. "An Innovation-Focused Scenario Process – A Case from the Materials Producing Industry." *Technological Forecasting and Social Change* 80 (4): 599–610.

West, J., and M. Bogers. 2014. "Leveraging External Sources of Innovation: A Review of Research on Open Innovation." *Journal of Product Innovation Management* 31 (4): 814–831.

West, J., and S. Gallagher. 2006. "Challenges of Open Innovation: The Paradox of Firm Investment in Open-Source Software." *R&D Management* 36 (3): 319–331.

Yang, H. L., and T. C. T. Wu. 2010. "Linking Technology Intelligence to Open Innovation." *Technological Forecasting and Social Change* 77 (2): 335–343.

The role of corporate foresight in exploring new markets – evidence from 3 case studies in the BOP markets

Jakob Højland and René Rohrbeck

ABSTRACT

We examine to what extent successful business-development activities in uncertain environments can be classified as corporate foresight (CF) and to what extent they have been intentional and systematic. Beginning with identifying successful cases of new-business development in Bottom of the Pyramid markets, we use various data sources to reconstruct timelines and map CF activities. We selected the cases to maximise their heterogeneity in firm size, industry, nature of the product and ownership structure. Our findings suggest that the probing (experimental search) phase is particularly important in unknown and uncertain environments but that perceiving and prospecting (cognitive search) activities are necessary to find distant opportunities. In addition, we find that successful business-development activities rely on multiple iterations between perceiving, prospecting and probing. Our findings emphasise that CF should include activities that encompass both experimental and cognitive search elements.

1. Introduction

Over the last decade, technological development such as digitalisation and the Internet of Things (IoT) have created novel opportunity spaces that are now being explored by firms to develop attractive new markets. Simultaneously, these developments disrupt established industries in which emerging technologies alter the existing rules of the game and enable entirely new business models (Johnson, Christensen, and Kagermann 2008). The increased speed and complexity of today's business environment calls for novel approaches that permit the anticipation and interpretation of trends and the use of the insights gained for a competitive advantage (Tsoukas and Shepherd 2004; Coates, Durance, and Godet 2010). In the past two decades, an increasing number of companies have built corporate foresight (CF) capabilities (Daheim and Uerz 2008). During this time, CF practices have evolved from specific deployment of methods and processes to being organisationally integrated and becoming a key component of the future preparedness of organisations (Rohrbeck, Battistella, and Huizingh 2015).

CF practices have been shown to be an important predictor for superior profitability and market valuation growth (Rohrbeck, Kum, and Jissink 2017). Through systematic CF practices, firms can overcome cognitive and action-level inertia (Gavetti 2012). CF practices permit the identification of superior courses of action and foreseeing their consequences (Gavetti and Menon 2016), which can help on the strategy level as well as on the level of innovating new products, services, and business models (Rohrbeck and Gemünden 2011; Heger and Rohrbeck 2012). While CF has been shown to create value in saturated Western markets (Rohrbeck 2012; Rohrbeck and Schwarz 2013), there are no similar studies that replicate the finding in developing markets.

In this paper, we investigate three cases of successful market exploration in a particularly challenging segment characterised by a high number of consumers with very low spending power. Such markets have become known as the Bottom of the Pyramid (BOP) segment (Prahalad and Hart 2002).

In the following we will first discuss the literature and introduce our analytical framework. In section 3 we explain our research approach. In section 4 we first analyse our three cases using timelines and our analytical framework and then report on the cross-case analysis. In section 5 we discuss our findings and discuss future research directions.

2. Building a framework for corporate foresight in BOP markets

2.1. Corporate foresight

CF has been regarded as an organisational ability, set of practices and process (Slaughter 1997; Becker 2002; Rohrbeck 2010a; Hines and Bischop 2006). In this paper, we conceptualise CF as a set of practices that support new business exploration through identifying drivers of change (perceiving), interpreting the consequences of the drivers both individually and collectively (prospecting), and actions that permit the testing of value propositions, product/service designs and market acceptance (probing).

Perceiving often starts with the acceptance that firms need to build specialised sensors that reduce blind spots in their peripheral vision (Day and Schoemaker 2004; Winter 2004). The aim is to identify the majority, if not all, of relevant change drivers in a given market space and gain a lead-time advantage towards competitors (Rohrbeck 2010b). The outcome of the perceiving process is a list of weak and strong signals of change, trends, critical events and strategic surprises (Peter and Jarratt 2015) rated by relevance for the firm. Often, firms will also organise them on dimensions such as the level of uncertainty, impact, maturity or domains. In that respect, we define uncertainty as the inability to anticipate the likely evolution of drivers of change in the business environment, which can be reduced to a certain extent though structured CF practices (Vecchiato and Roveda 2010).

Prospecting includes practices that translate signals into insights. Insights are created through interpreting the signal and the consequences for the focal firm. In new market exploration, a particular focus is on identifying systemic effects, in which multiple drivers are interdependent, mutually reinforcing and able to trigger exponential change. This phenomenon has previously been described as the S-Curve and is today also often called the tipping point. The tipping point is the moment that a product or service becomes mainstream, is publicly accepted and sees exponential growth in demand (Gladwell 2001). Prospecting involves the usage of structured methods such as scenarios, roadmaps or systems-dynamics analysis to anticipate potential and plausible development trajectories that can inform the decision on the course of actions (Gavetti and Menon 2016). Prospecting practices allow the firm to reduce the effect of uncertainty, i.e. the lack of knowledge about the consequences of a change. Furthermore, it is the basis for reducing response uncertainty, which is the inability to identify viable responses in a given situation and understand the consequences of the choice (Vecchiato and Roveda 2010). Ultimately, the response is shaped by the bounded rationality of decision makers, as their capability to make rational decisions is limited by past experiences and current knowledge (Vecchiato 2015).

Probing practices are designed to trigger meaningful organisational responses to change. Probing is the first phase of direct action. This action can be a direct response to the insights generated in the perceiving and prospecting phases, or it can be part of an experimental search scheme in which firms experiment to gain insights (Gavetti and Levinthal 2000). Examples of probing are initial prototyping, R&D projects, establishments of strategic partnerships and consumer tests with novel products. The nature of these actions differs across industries. Increasingly, firms are also using external venturing schemes to experiment outside of the firms' boundaries or through internal venturing, in which teams of the firms' own employees are supported in building a new solutions, products or businesses (Rohrbeck, Döhler, and Arnold 2009). Probing is an action-based approach that complements the practices of perceiving and prospecting.

While the phases are building on each other, they do not necessarily occur in a linear sequence; rather, they are complementary phases that can form part of an iterative mechanism that translates signals into insights and action. Firms that regularly iterate between phases will profit from powerful learning loops (Gavetti and Levinthal 2000).

2.2. Bottom of the pyramid

The reduction of market protection mechanisms and the consequent openings of economies around the globe have radically transformed the way business is done today (Prahalad and Hart 2002). Firms are facing challenges in the form of increased competition and slowing growth in developed markets, which has forced businesses to be creative and look into the potential in new and less-developed markets in the global economy (Prahalad and Hart 2002). Still, few take the leap and realise the potential for the growing consumer segment of the world's more than 4 billion people living for $2–4 a day, which is known as the BOP segment (Dansk Industri 2007).

For our analysis, we define three individual paths for developing BOP business opportunities.

Building a Market through Local Co-Creation and Deep Dialogue emphasises that in the BOP segments, markets need to be built together with the consumers. According to Prahalad and Hart (2002), this often includes the need to provide access to credit and increase the earning potential of the poor. Another method to increase purchasing power is to introduce pay-as-you-go business models or single-serve packages to address the buying power issue of the BOP segment (Prahalad and Hart 2002; Dansk Industri 2007). Simanis and Hart (2008) suggest a strategy called the BOP protocol, the first step of which is to open the market and overcome barriers through deep dialogue with local stakeholders to increase the firm's credibility and transparency in the new local market.

Creating a Sustainable Value and Distribution Chain emphasises the creation of a sustainable business model and achieving scale in all parts of the value chain from production to distribution to ensure the long-term viability of the business model (Dansk Industri 2007). When scaling in all parts of the value chain, it is possible for the focal firm to drive down costs to accommodate the available funds of the BOP consumer (Dansk Industri 2007). Prahalad and Hart (2002) argue that firms need to invest and promote sustainable development beginning with the BOP segment. This builds credibility and will later have great influence on the consumers in Tier 1, i.e. the most financially affluent in society. Furthermore, it is important to improve access to the BOP, which is geographically dispersed to be sustainable in distribution networks and communication across geographical areas (Prahalad and Hart 2002). Therefore, MNCs either need to be creative and establish ties directly to the end consumer or, more likely, partner with a local champion in the BOP market to navigate through the complexity of the distribution and value chain.

Create Tailored Product Offerings Profitably emphasises that firms should be able to adapt their product offerings to accommodate the tastes and budgets of local consumers. According to Prahalad and Hart (2002), this product adaptation to local markets and cultures should be driven through collaboration with local consumers. In that respect, Simanis and Hart (2008) argue that this is the time for the company to move into the enterprise 'creation mode', which is where the profitable business is created, i.e. not through simply expanding their existing offerings to new regional markets. This adaptation is also key to achieving scale in the value and distribution chain, which can be expected as a prerequisite for the firm to achieve profitability on their products.

Closing the loop on this BOP section and the main topic of CF, we argue that the level and structure of CF activities in firms create the foundation for the BOP business-development phase, as illustrated in our conceptual framework in Figure 1.

We employ this *conceptual model* as a research lens through which we observe and discuss the actual business-development sequence.

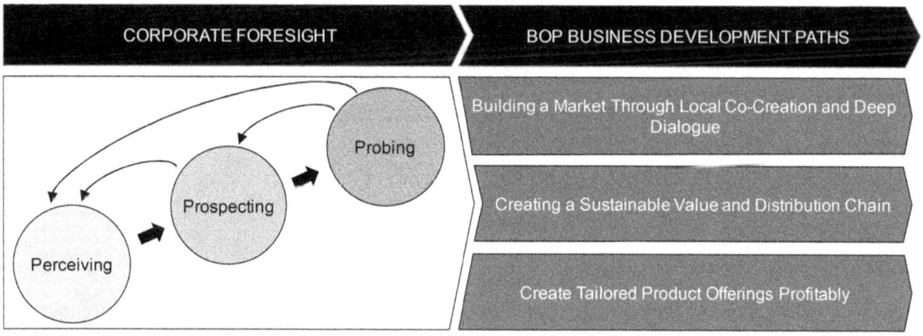

Figure 1. Conceptual model.

3. Research approach

In our research, we investigate the question to which role deliberate & structured vs. unstructured CF play in the exploration of new markets in uncertain environments. We use CF practices and the BOP-paths frameworks as guides for deductive reasoning but enter the field with open interview guides and openness towards uncovering alternative explanations (Saunders, Lewis, and Thornhill 2007).

3.1. Research setting and case selection

In our study we have applied a maximum-variation criteria to identify cases which were different on size, industry and ownership structure, see Table 1 (Eisenhardt 1989). The maximum variation also boosts the generalizability of our findings beyond the context of a single firm (Eriksson and Kovalainen 2008; Flyvbjerg 2006; Yin 2003).

In addition, we tested that the cases meet three criteria: (a) benefitting the BOP segment, (b) operating in Africa, and (c) the accessibility of key employees involved in the process of developing the business.

3.2. Data collection, reduction and coding

For our data collection, we built on secondary data from reports and internal documents in addition to face-to-face and phone interviews (Rubin and Rubin 2005). For the interview, we relied on semi-structured questionnaires and the draft timelines that we had prepared on the basis of the secondary data. In total, eight interviews were conducted: two from each case company, which included always the main responsible for the business development activity and one team member. In addition, we interviewed for triangulation one international expert on the BOP segment and one Danish business-in-Africa experts.

Table 1. Overview cases.

	Evershelter	Kamstrup	Arla
Industry	Aid & relief	Engineering	Food & beverage
Industry characteristics	Heavily influenced by regulation	Reliance on high-tech components	High interdependencies between product and production process
Nature of product	New to the world Market pull Built for BOP market	New to Africa Technology push Adapted to BOP market	New to Africa Market pull Adapted to BOP market
Size	Small, start-up	Mid-sized, 1000 employees	Large, 18,000 employees
Ownership structure	Privately owned	Owned by large energy company	Cooperative

The processing of the data followed in the first step deductive principles using the key constructs of CF and the BOP paths as a form of thematic coding (Eriksson and Kovalainen 2008). In the second round of coding, we manually reviewed all transcripts to identify additional areas of interest. This coding was driven by more openness towards new patterns and emergent findings using inductive reasoning (Eisenhardt 1989). We applied this two-phase data-reduction and coding process first to the within-case level and then to our cross-case analysis.

4. Results

In the results section, we will first present our within-case findings for all three cases. In section 4.4 we then discuss the extent to which the practices have contributed to the success of the business-development activity. Special attention is placed on market entry and business-model adaptation. Where we conceptualise 'business model', as the architectural setup a firm uses to define a compelling value proposition and create and capture value (Johnson, Christensen, and Kagermann 2008).

4.1. Start-up company in housing industry (Evershelter)

Our first case, Evershelter, is a start-up that produces shelter solutions for displaced people. The firm operates in the aid-and-relief industry, serving the BOP segment by providing an alternative to other temporary housing solutions, which are primarily tents.

Figure 2 shows the sequence of the actual business-development activities. It can be seen that the early perceiving (step 1) was the result of an unsystematic activity. The initial idea was triggered by a long-term housing project for refugees for which the father of the CEO of Evershelter (JC) was responsible. JC explained that he initially became interested when he saw reports about refugees. He then realised that technology today can enable alternative solutions to tents that would be 'more robust and which could hold longer [...]'. This was particularly relevant, as refugees stayed increasingly longer in camps (at that time an average of 15 years) that were designed to be short-term shelters and not long-term housing. At that stage, JC was still acting alone and started to perform his own prospecting (step 2), beginning with technology scanning and looking on the Internet for alternatives to tents and mapping possible technical solutions. On that quest, he uncovered a potential technology provider in Denmark and moved on to investigate the market. To further explore the viability of

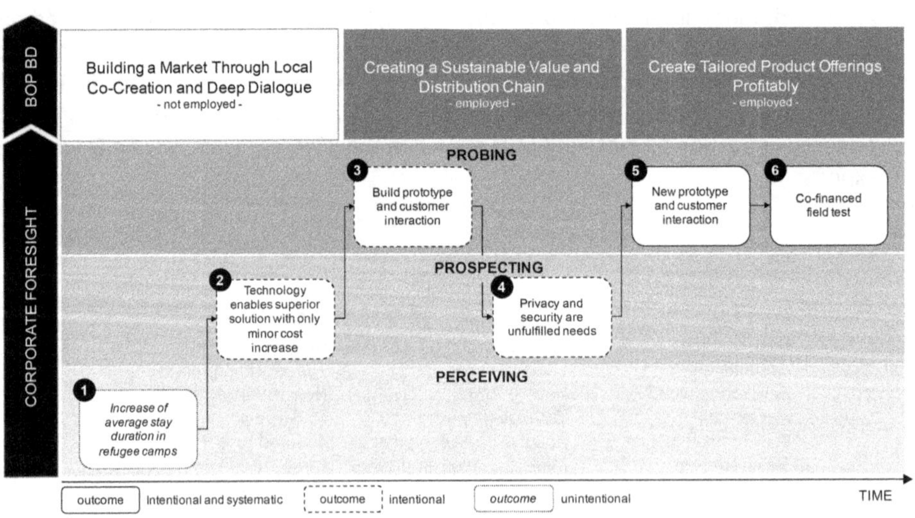

Figure 2. Business development sequence – Evershelter.

the business opportunity, JC reviewed refugee statistics from the UN, finding that the number of people living in refugee camps was starting to grow dramatically, increasing his hope of doing good and simultaneously providing a viable business opportunity. For probing (step 3), JC used the opportunity that he was enrolled in a master's course on disaster relieve, in which the other participants were all 'experienced camp people, who had flown in from around the world [...] and I just got a lot of great feedback'.

With encouraging feedback on the first prototype, JC was able to find a first innovation incubator that provided 60% of the investment for building the first prototype. At the same time, he found with Real Relief a complementary business partner that was headed by a well-respected and innovative business woman from the aid-and-relief industry, according to one of the interviewed experts on Africa. This partnership brought significantly more insight into the shelter business and consequently helped create through prospecting (step 4) a deeper understanding of the future business model and the elements that would be crucial for a solution and commercial success. Among these, it was discovered that the tents lacked privacy and security, which are particularly important for long-term residents. With this reinforced understanding, the second round of probing (step 5) was prepared with the development of a more stable prototype. With this prototype, they engaged in field tests, which included one unit in Iraq that subsequently was expanded to 33 and 14 in Uganda. This probing was made possible through the 'Dansk Flygtningehjælp'. This second probing phase also led to two major product improvements: making the doors lockable and allowing them to let more light through. Finally, in the third probing phase (step 6), further improvements were investigated, including designs to make the shelter earthquake proof. This probing was effectuated by involving the Universities of Aarhus and Copenhagen, which also allowed the limiting of costs and ensured the use of state-of-the-art technology.

With regards to the BOP business-development activities, we attest that Evershelter has mostly used two of the three paths.

- They embraced the *creating a sustainable value and distribution chain* through engaging in the partnership with Real Relief and with the Danske Flygtningehjælp to build a value chain reaching all the way to the end customer.
- They also worked on *creating tailored product offerings profitably* by starting their development with the end customer with the BOP needs (security and privacy) in mind.

When looking at how intentional and systematic the practices were, we can find a typical start-up pattern in which the initial value proposition idea is not derived from a systematic approach but from a person-specific insight that was picked up unintentionally. In the later stages, one could have expected that the more the company needs to rely on partnerships and external funding, the more it would need to engage in more structured prospecting activities. Here, however, the approach was still heavily instinct driven, which the head of Real Relief commented on as a potential reason for missing valuable opportunities. However, the last two phases were both intentional and systematic.

4.2. Multinational corporation in the engineering industry (Kamstrup)

Our second case, Kamstrup, is a world-leading company within the industry of intelligent metering solutions, operating in 24 countries worldwide as a direct supplier for energy and utility companies. Since 2009, Kamstrup has served the African market, including the BOP segment.

In the case of Kamstrup, we would expect that, as an established firm owned by an even larger mother company, it would have a more structured approach to new business development than our first start-up case. Indeed, the perceiving activities are following an intentional and systematic practice that is the business-development team drives. The practice consists of using external networks and reports from think tanks and media-monitoring data. Of the perceiving function, key elements are outsourced to external consultants and futurists.

Figure 3 gives an overview of the business-development activities. As a result of the initial perceiving activity (step 1) several mega trends were identified, one of which was the lack of access to electricity in developing countries. In many countries, this trend had already triggered meaningful governmental and NGO activities. In some parts of Africa, the share of consumers with access to electricity is as low as 30%. Consequently, many African utility companies are actively seeking smart energy management solutions, which triggered probing activities (step 2) in Kamstrup. These probing activities consist of product adaptations to the local needs. Whereas these probing activities ensured that the solution-market fit was enhanced, it was not sufficient to enter into large-scale contracts, and the business-development activities in Africa moved back to the prospecting phase (step 3). The prospecting effort was aimed at predicting and managing the tipping point, which would pave the way for large-scale adoption of smart energy solutions.

Through working with the local customers, Kamstrup could substantiate that a tipping point is approaching. Consequently, they moved to the next probing phase (step 4) towards acquiring the necessary certificates for a number of African countries. With the certificates in hand, they could reengage with the customer, with whom they had already, during the prospecting phase, built a deep relationship, and move towards the proof-of-concept (PoC) phase (step 3). Such PoCs include a 200-device pilot project in Kenya. To invest in large-scale deployments, these probing activities are essential; as the head of sales explains:

> Here you get to see reality and know the customer, the environment of the metering device, and pressure test your product to make sure that it can perform, so you will also increase customer confidence that the product will solve his business issue.

From the BOP business-development paths,

- The *creation of tailored solutions* through the PoC phase was particularly pronounced. The product needed to be adapted to African needs, and Kamstrup worked towards that goal directly with the local utilities. This work was needed, as the customers in Africa were faced with very different challenges and thus needed different value propositions, e.g. increased access to energy in Africa vs. increased energy efficiency in Europe.

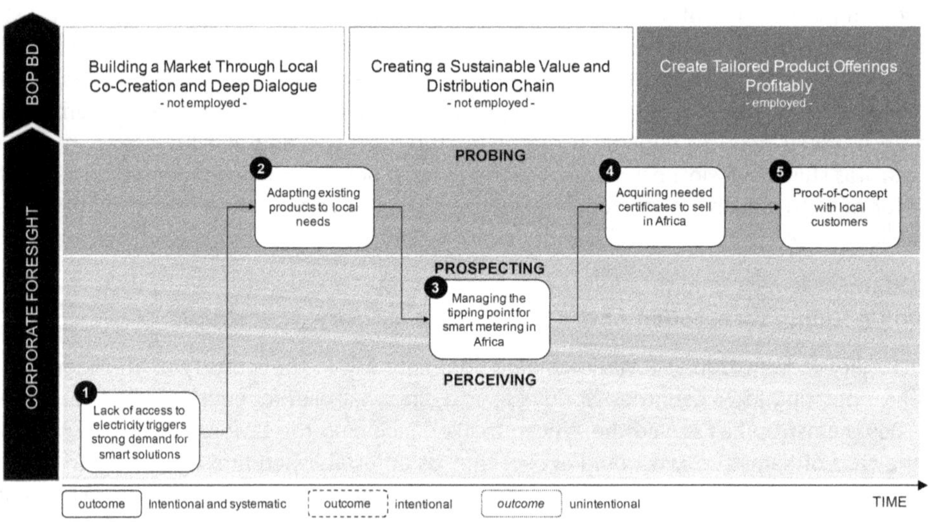

Figure 3. Business development sequence – Kamstrup.

4.3. Multinational corporation in food & beverages industry (Arla)

Our last case, Arla, is a global dairy company. Their products are fast-moving consumer goods, for which prices are low but volumes are high. Arla operates in markets that need high degrees of product adaptations to the specific needs and tastes of the local consumer groups. Figure 4 shows the business-development sequence that led to this market success.

The perception phase (step 1) of this success story begins with a lack of perception of a strategic surprise, i.e. the weak signals have remained undetected, and the strategic surprise was only spotted in the fait-accompli state. This strategic surprise was the removal of subsidies that were the basis for the profitable business case of shipping European milk to Africa. The logical consequence would have been to pull out of the market as quickly as possible. Through intentional, but at the start, mostly unstructured perceiving (step 2) Arla identified two key trends: the growing middle class and the removal of trade barriers, which could eventually lead towards a very attractive future market. However, it was also clear that in the short-term, a solution needed to be found to drastically reduce production costs. This solution was found through prospecting (step 3). As a first insight, it became clear that a new solution would need to meet three criteria to be considered a viable BOP product: (i) it needs to have a 'long life' because otherwise it could not be produced on the basis of European milk and shipped to Africa, (ii) it needs to be 'ambient', i.e. it has to be possible to store and distribute without cooling and (iii) it needs to be 'affordable'. The solution was milk powder, which had the same nutritional qualities as whole milk powder but in which the milk fat was substituted by vegetable fat, which drastically lowered the production cost.

The first probing phase (step 4) was thus mainly concerned with developing the right formula for the milk and developing the production process. The second phase of probing (step 5) was prototyping the production to start going through the official approval tests to gain market access. After gaining the necessary certificates, the market development went smoothly, and Arla became a market leader in a number of regions in Nigeria. However, unexpected competitor lawsuits resulted in an injunction that prevented Arla from selling its product. Through an additional prospecting effort (step 6), it was possible to adapt the design and circumvent the injunction.

Another interesting aspect in the case was the comprehensive application of the BOP business-development paths.

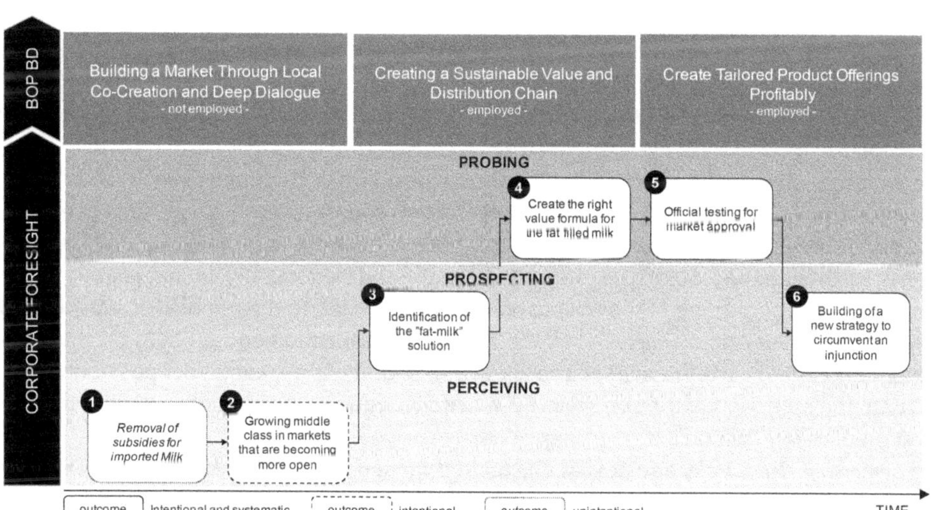

Figure 4. Business development sequence – Arla.

- Arla has been successful in *building the market through local co-creation*, in particular through multiple rounds of consumer tests in Africa. Here, Arla investigated and cocreated product characteristics, such as taste, size and price point.
- Arla had earlier understood the need to *tailor its offerings*, leading to the low-cost, fat-filled formula. In addition, effort also went into adapting the milk's taste to local preferences.
- More importantly, they understood that large-quantity packaging could be an to buy the product. This insight led to the development of 6-g single-serving packages that worked well for consumers in BOP markets. However, another challenge was that this packaging and the associated distribution system would become too expensive if the packaging needed to occur in Europe. A solution was to build a local packaging site that formed the basis for a *local value and distribution chain*. This solution allowed positioning the product in the low-cost market and, as an additional benefit, led to local job creation, which was seen as a highly valued local engagement that in turn boosted the credibility of Arla as a firm and the brand attachment.

4.4. Cross-case analysis

In the cross-case analysis we first looked for patters by systematically going through the firm characteristics. When looking for effects from *industry* on CF practices, theory would expect to find more intentional and systematic practices in industries prone to disruptions (Müller and Müller-Stewens 2008; Sarpong and O'Regan 2014). In our case, Kamstrup, from an industry with a high level of disruption, has indeed been more active than Arla, which operates in a comparably low-disruptive environment. Evershelter operates in an industry that is highly regulated and dependent on political influences. The initial idea and a great amount of the prospecting has consequently been focused on the political level, relying, for example, on UN reports. The higher deployment of systematic foresight techniques can, in the case of Kamstrup, also have resulted from the need for heavy investments in technology development and systems integration, which translates into pressure to seek economies of scale by expanding globally.

When looking for effects from the *nature of the product*, we would expect that technology-push products would be complemented by market perceiving and prospecting activities (Hofmann 2015), which was also the case for Kamstrup. For market-pull products, dedicated scouting for technology would be expected. In Evershelter, that also occurred, while in the case of Arla, it was more a matter of using internal technological capabilities to build the new BOP product. We also found that the products that existed and were then adapted to Africa relied more on probing (experimental search) and less on prospecting (cognitive search). However, this might also reduce their impact, as their positioning might be closer to the current products than what cognitive search might have revealed.

The expected effects of *size* would be that the larger a firm is, the more likely it would be to have established formal CF capabilities (Schwarz 2008; Daheim and Uerz 2008). Large firms need to protect assets and ensure that the firm can renew its competitive advantage (Ruff 2015; Boe-Lillegraven and Monterde 2015; Hines and Gold 2015; Rohrbeck, Arnold, and Heuer 2007). In that respect, it comes as a surprise that the monitoring of the political situation, which led to the removal of subsidies, was not perceived early in the Arla case. However, the following prospecting and probing mechanisms were still potent enough to save the market and allow Arla to attain a market-leader position with its new product. On the other end of the size scale, Evershelter conducted its foresight activities in a more ad-hoc and on-demand fashion.

The *ownership structure* is a relatively unexplored influencing factor. However, we enter the field with the expectation that privately or cooperatively owned companies are more likely to take a long-term investment perspective than shareholder-owned firms, and consequently, that they would be more interested in investing in CF (van der Duin and den Hartigh 2009). In our case, we attest that the influence of ownership structure is weak and overridden by other factors. In particular, in the comparison of Arla and Kamstrup, we see that the nature of the product has mostly led to Arla being less invested in structured foresight approaches than Kamstrup.

In addition to the influence of firm characteristics, we wanted to search for patterns in the business-development sequence. Here the first notable commonality was that all firms underwent iterations within the 3 Ps (perceiving, prospecting, and probing) of CF. The implication is that firms need to stay persistent when exploring new markets with foresight and be ready to iterate between thinking (perceiving, prospecting) and acting (probing). Explained differently, they can profit from feedback loops between cognitive and experimental search activities.

To differentiate the extent to which the actual business-development activities match with the prescriptions of CF practices, we introduced the following levels: a) unintentional (no formal CF activities and no planned, deliberate effort), intentional (still no formal CF practices but planned and deliberate) and intentional and systematic (purposefully exercised CF practice). We find that CF practices were only in one out of three cases the trigger for the business-development activity. In all cases, we find intended and systematic CF activities somewhere in the business-development sequence (typically in the second half of the sequence). This finding suggests firms' weakness to repeatedly and systematically trigger new business-development. It also shows that once triggered, the firms can rely more on established and effective CF methods and tools. This finding resonates well with the suggestion of the innovation's fuzzy-front-end literature that the main differentiator between average and outperforming firms might be in the degree of mastery of the early and initial phases (Rohrbeck 2013; Price et al. 2009).

In addition, we find that when complexity and uncertainty is high, firms tend to engage in early and extensive probing (experimental search) activities. This is insofar justified that CF requires a certain level of predetermination, i.e. justified expectations about the direction and rate of change (Van der Heijden 2005). However, it carries the risk that the identified opportunities are not sufficiently distant to the current market to carry superior value creation potential. As Gavetti and Levinthal (2000) emphasise, cognitive search (perceiving and prospecting) is more potent than experimental search to uncover distant alternatives.

5. Discussion

The aim of our study was to investigate the extent to which actual and successful business-development activities can be classified as CF and the extent to which they have been intentional and systematic. We placed our empirical investigation in BOP markets deliberately for two reasons. First, we see these markets as being particularly relevant for creating value for firms and societies alike. Second, we use them as an environment with a high level of uncertainty and as markets where firms have very limited past experience.

To boost generalizability, we selected cases that were very different from each other, which limited our ability to use replication for the validation of our findings. Our findings are thus more indicative than conclusive. The phenomena we found and discussed should be thus subject to further analysis.

Our research approach is heavily reliant on timelines, which we constructed through secondary data (internal and external reports and documents), actor interviews (with two respondents per case) and interviews with external observers and Africa experts. This allowed us to ensure a moderate level of triangulation, which reduced a potential single-source bias. However, the retrospect bias could have negatively influenced our data. Future studies could overcome these limitations by using more data-collection techniques, such as observations, diaries and more detailed documents, such as minutes of key meetings, project presentations and email-based discussions.

6. Conclusion

With this article, we want to contribute to our understanding of the impact of CF activities. While research on CF has typically taken its starting point in observing intended and systematic approaches

(Ruff 2006; Peter and Jarratt 2015), we started from the other direction. This provides us with an interesting vantage point from which we can systematically challenge the wisdom that CF activities contribute to the exploration success of a firm.

We find that in our sample, firms were using systematic CF methods only rarely in the early phases, suggesting that there is a high risk that opportunities remain unidentified and consequently unexplored and unexploited. In the latter phases, however, more systematic activities are conducted that include scanning for and matching of technology and market changes and systematically constructing value formulas and business models.

We further found that the successful cases went through multiple iterations of perceiving, prospecting and probing, suggesting that successful business development is a non-linear process that builds on feedback loops and requires persistence. For theory our finding suggests that cognitive and experimental search mechanisms must be combined to create effective exploration capabilities (Gavetti and Levinthal 2000). This could permit overcoming the weaknesses of the one search mode, e.g. the inability to identify distant opportunities with experimental search, with the strengths of the other search mode, e.g. the ability to use cognitive search to overcome the experience bound.

The setting in the BOP markets placed us empirically in a high uncertainty environment, which forms a stark contrast towards business development in developed countries and markets, in which, for most, driving factors have a high level of predetermination and data availability is high. In this context, we can attest that firms relied heavily on early and extensive probing activities. These experimental search activities seemed to have a dual role. First, they provided tangible POCs, which aid in decision making. Second, they allowed the generation of insights into market and technology aspects that would ex-ante not have been judged to be relevant.

In conclusion, we expect that also outside our investigation scope, the BOP markets, the importance of probing activities has been underestimated. While traditional CF literature prescribes many methods, tools and practices for cognitive search (perceiving and prospecting), we conclude that in future investigations, we need to pay more attention to the need for and benefits of experimental search (probing).

Disclosure statement

No potential conflict of interest was reported by the authors.

ORCID

René Rohrbeck http://orcid.org/0000-0002-2320-7915

References

Becker, Patrick. 2002. "Corporate Foresight in Europe: A First Overview." In *Working Paper European Commission*, 31. Brussels: European Commission.

Boe-Lillegraven, Siri, and Stephan Monterde. 2015. "Exploring the Cognitive Value of Technology Foresight: The Case of the Cisco Technology Radar." *Technological Forecasting and Social Change* 101: 62–82. doi:10.1016/j.techfore.2014.07.014.

Coates, Joseph, Philippe Durance, and Michel Godet. 2010. "Strategic Foresight Issue: Introduction." *Technological Forecasting and Social Change* 77 (9): 1423–1425. doi:10.1016/j.techfore.2010.08.001.

Daheim, Cornelia, and Gereon Uerz. 2008. "Corporate Foresight in Europe: From Trend Based Logics to Open Foresight." *Technology Analysis & Strategic Management* 20 (3): 321–336.

Dansk Industri. 2007. *Working with the BOP-Success in Low Income Markets*. Copenhagen, Denmark: Dansk Industri.

Day, George S., and Paul J. H. Schoemaker. 2004. "Driving Through the Fog: Managing at the Edge." *Long Range Planning* 37 (2): 127–142.

Eisenhardt, K. M. 1989. "Building Theories from Case-study Research." *Academy of Management Review* 14 (4): 532–550.

Eriksson, P., and A. Kovalainen. 2008. *Qualitative Methods in Business Research*. London, UK: SAGE.

Flyvbjerg, B. 2006. "Five Misunderstandings About Case-study Research." *Qualitative Inquiry* 12 (2): 219–245. doi:10.1177/1077800405284363.

Gavetti, G. 2012. "PERSPECTIVE—Toward a Behavioral Theory of Strategy." *Organization Science* 23 (1): 267–285. doi:10.1287/orsc.1110.0644.

Gavetti, Giovanni, and Daniel Levinthal. 2000. "Looking Forward and Looking Backward: Cognitive and Experiential Search." *Administrative Science Quarterly* 45 (1): 113–137.

Gavetti, Giovanni, and Anoop Menon. 2016. "Evolution Cum Agency: Towards a Model of Strategic Foresight." *Strategy Science* 1 (3): 207–233.

Gladwell, Malcolm. 2001. *The Tipping Point: How Little Things Can Make a Big Difference*. New York: Little, Brown.

Heger, Tobias, and René Rohrbeck. 2012. "Strategic Foresight for Collaborative Exploration of New Business Fields." *Technological Forecasting and Social Change* 79 (5): 819–831.

Hines, Andy, and Peter Bischop. 2006. *Thinking About the Future: Guidelines for Strategic Foresight*. Washington, DC: Social Technologies.

Hines, Andy, and Jeff Gold. 2015. "An Organizational Futurist Role for Integrating Foresight into Corporations." *Technological Forecasting and Social Change* 101: 99–111. doi:10.1016/j.techfore.2014.04.00.

Hofmann, Rupert. 2015. "Visionary Competence for Long-term Development of Brands, Products, and Services: The Trend Receiver Concept and its First Applications at Audi." *Technological Forecasting and Social Change* 101: 83–98. doi:10.1016/j.techfore.2014.06.005.

Johnson, Mark W., Clayton M. Christensen, and Henning Kagermann. 2008. "Reinventing Your Business Model." *Harvard Business Review* 86 (12): 57–68.

Müller, Adrian W., and Günter Müller-Stewens. 2008. *Strategic Foresight: Trend- und Zukunftsforschung in Unternehmen-Instrumente, Prozesse, Fallstudien*. Stuttgart: Schäffer-Poeschel.

Peter, Marc K., and Denise G. Jarratt. 2015. "The Practice of Foresight in Long-term Planning." *Technological Forecasting and Social Change* 101: 49–61. doi:10.1016/j.techfore.2013.12.004.

Prahalad, Coimbatore K, and Stuart L Hart. 2002. "The Fortune at the Bottom of the Pyramid." *Strategy and Business* 26: 2–14.

Price, R. L., A. Griffin, B. A. Vojak, and N. Hoffmann. 2009. "Innovation Politics: How Serial Innovators Gain Organisational Acceptance for Breakthrough New Products." *International Journal of Technology Marketing* 4 (2): 165–184.

Rohrbeck, René. 2010a. *Corporate Foresight: Towards a Maturity Model for the Future Orientation of a Firm*. Heidelberg: Physica-Verlag, Springer.

Rohrbeck, René. 2010b. "Harnessing a Network of Experts for Competitive Advantage-Technology Scouting in the ICT Industry." *R & D Management* 40 (2): 169–180.

Rohrbeck, René. 2012. "Exploring Value Creation from Corporate-foresight Activities." *Futures* 44 (5): 440–452.

Rohrbeck, René. 2013. "Trend Scanning, Scouting and Foresight Techniques." In *Front End of Innovation: Managing the Unmanagable Fuzzy Side*, edited by Fiona Schweitzer and Oliver Gassmann, 59–73. Heidelberg, Berlin: Springer.

Rohrbeck, René, Heinrich Arnold, and Jörg Heuer. 2007. "Strategic Foresight in Multinational Enterprises-a Case Study on the Deutsche Telekom Laboratories." Paper presented at the ISPIM-Asia 2007 conference.

Rohrbeck, René, Cinzia Battistella, and Eelko Huizingh. 2015. "Corporate Foresight: An Emerging Field with a Rich Tradition." *Technological Forecasting and Social Change* 101: 1–9. doi:10.1016/j.techfore.2015.11.002.

Rohrbeck, René, Mario Döhler, and Heinrich Arnold. 2009. "Creating Growth with Externalization of R&D Results-the Spin-along Approach." *Global Business and Organizational Excellence* 28 (4): 44–51.

Rohrbeck, Rene, and Hans Georg Gemünden. 2011. "Corporate Foresight: Its Three Roles in Enhancing the Innovation Capacity of a Firm." *Technological Forecasting and Social Change* 78 (2): 231–243.

Rohrbeck, Rene, Menes Etingue Kum, and Tymen Jissink. 2017. *Corporate Foresight: Benchmarking Report 2017*. Aarhus, Denmark: Aarhus University, Aarhus School of Business and Social Sciences.

Rohrbeck, René, and Jan Oliver Schwarz. 2013. "The Value Contribution of Strategic Foresight: Insights from an Empirical Study of Large European Companies." *Technological Forecasting and Social Change* 80 (8): 1593–1606.

Rubin, H. J., and I. Rubin. 2005. *Qualitative Interviewing: The Art of Hearing Data*. Thousand Oaks, CA: Sage.

Ruff, Frank. 2006. "Corporate Foresight: Integrating the Future Business Environment into Innovation and Strategy." *International Journal of Technology Management* 34 (3-4): 278–295.

Ruff, Frank. 2015. "The Advanced Role of Corporate Foresight in Innovation and Strategic Management-reflections on Practical Experiences from the Automotive Industry." *Technological Forecasting and Social Change* 101: 37–48. doi:10.1016/j.techfore.2014.07.013.

Sarpong, David, and Nicholas O'Regan. 2014. "The Organizing Dimensions of Strategic Foresight in High-velocity Environments." *Strategic Change* 23 (3-4): 125–132. doi:10.1002/jsc.1965.

Saunders, Mark, Philip Lewis, and Adrian Thornhill. 2007. *Research Methods for Business Students*. 4th ed. Harlow, England: New York: Financial Times/Prentice Hall.

Schwarz, Jan Oliver. 2008. "Assessing the Future of Futures Studies in Management." *Futures* 40 (3): 237–246.

Simanis, E., and S. L. Hart. 2008. *The Base of the Pyramid Protocol-towards Next Generation BOP Strategy*. Ithaca, NY: Cornell University.

Slaughter, Richard A. 1997. "Developing and Applying Strategic Foresight." *ABN Report* 5 (10): 13–27.

Tsoukas, H., and J. Shepherd. 2004. "Coping with the Future: Developing Organizational Foresightfulness-introduction." *Futures* 36 (2): 137–144.

van der Duin, Patrick, and E. den Hartigh. 2009. "Keeping the Balance. Exploring the Link of Futures Research with Innovation and Strategy Processes." *Technology Analysis & Strategic Management* 21 (3): 333–351.

Van der Heijden, Kees. 2005. *Scenarios: The Art of Strategic Conversation*. 2nd ed. Chichester, West Sussex: John Wiley & Sons.

Vecchiato, Riccardo. 2015. "Creating Value Through Foresight: First Mover Advantages and Strategic Agility." *Technological Forecasting and Social Change* 101 (25-36). doi:10.1016/j.techfore.2014.08.016.

Vecchiato, Riccardo, and Claudio Roveda. 2010. "Strategic Foresight in Corporate Organizations: Handling the Effect and Response Uncertainty of Technology and Social Drivers of Change." *Technological Forecasting and Social Change* 77 (9): 1527–1539.

Winter, Sidney G. 2004. "Specialised Perception, Selection, and Strategic Surprise: Learning from the Moths and Bees." *Long Range Planning* 37 (2): 163–169.

Yin, Robert K. 2003. *Case Study Research: Design and Methods*. 3rd ed. *Applied Social Research Methods Series*. Thousand Oaks, CA: Sage Publications.

Back to the future of women in technology: insights from understanding the shortage of women in innovation sectors for managing corporate foresight

Vinh Sum Chau and Chanell Quire

ABSTRACT

This paper investigates why there is a shortage of women in innovation, such as science, particularly technology, engineering and mathematics (STEM) sectors and offers insights for managing corporate foresight. It acknowledges that traditional corporate foresight methodologies have their own inherent problems, but argues that greater inclusion of women brings in new dimensions not previously recognised within the predominantly male-dominated technology sector. While extant feminist research may look at the general disadvantages women have in the workplace, few have examined the genesis and constitution of femininity to understand what new input can be brought innovation management, and how these different views can change the conduct of corporate foresight in the technology sector. Interviews from five senior personnel in the technology sector were conducted, and responses to a concise questionnaire involving 365 participants were obtained. Three case-rich narratives are presented as a summary on the future of women in technology.

1. Introduction

Improving corporate foresight methodologies for innovation management is part of the bigger picture for how they can be complemented by something that has long been lacking: the female perspective. This paper argues that the technology sector mostly, in which innovation management is key, has suffered a shortage of feminine involvement which constrains technological insights (such as in product development of female-specific products). In a world that is increasingly pushing for greater sex equality and recognition of a market for female specific IT products (eg. mobile phone apps for menstruation, pregnancy, breastfeeding, exercise and vitamin intake, etc), any corporate foresight activity that is limited in this way is likely to forego recognition of the market, as well as traits that women are more prone to having that may be useful in the conduct of corporate foresight exercises.

Hence, this paper goes backwards to examine the root of this problem, to suggest how the difference of women can offer insights for corporate foresight activities in innovation/technology that is not already understood in a male-dominated sector. To understand this, we (i) examine the reasons why many women are not desiring to pursue a career in technology, (ii) explore ways women can be encouraged to pursue technology-related careers, and (iii) identify the key attributes that constitute a woman that can advance innovation management. An inductive stance is adopted for the research that draws from women in technology management, while testing deductively some data collected from technology employment, and presenting case rich discourses on the implications for corporate foresight exercises. The context of the study is the UK, representing western development, rather than a global representation, as numerous such sensitive factors as culture, country employment laws, religion, etc, may otherwise distort the study, and are therefore assumed constant.

2. The need for women in corporate foresight exercises

The term 'corporate foresight' is used in this paper as a collective form to mean making meaningful predictions about future opportunities and threats for the organisation concerned, although Rohrbeck, Battistella, and Huizingh (2015) explained the evolution of the meaning of 'foresight', and how organisational foresight, managerial foresight, industrial foresight, strategic foresight and corporate foresight differ from each other. Earlier recommendation in the present Journal (van der Duin and den Hartigh 2009) has included the need of futures researchers to balance the way futures research is done to reflect the innovation and strategy processes of the organisation concerned. This means, if the sector is one that is opening up to a particular need or is attentive to an emerging market, the company's associated technological innovation must be both knowledgeable and empathetic of that new market; this in turn requires greater inclusion of women in the methodology of its corporate foresight exercise if that market is specifically women's technological products. A common criticism, for example, is that a recent Apple Corporation's iPhone health app was entirely based on the functions (and configuration) of men, and was initially launched without involvement, consultation with, or related foresight activities including, women (Lewis 2014; FT 2015). This has led to an immediate fall in its mobile phone market to a competitor that was perhaps more comprehensive in its corporate foresight methodology.

Hence, Massari Coelho et al. (2012) has called for more explicit application of foresight into the management plan of organisations and to identify the importance of this in science, technology and innovation sectors as a basis for global sustainable development. Specifically in the UK, recent statistics indicate a gender imbalance within STEM (science, technology, engineering and mathematics) sectors, and the technology sector stands out as being male-dominated (Wise Campaign 2015). This is concerning, as 72% of UK businesses rely on people with STEM skills while 58% of all new jobs will be STEM related, adding pressure to the need for women to become more involved (STEM Careers 2015). When new vacancies arise that are described as powerful, challenging and involves high levels of responsibility, they seem to appeal more to men than women; and despite the numerous developments and promotion of gender equality and diversity at work, there has only been around 15% increase in women in STEM occupations and around 35% increase in technology related ones in the recent few years (ibid).

Corporate foresight in innovation management stands out from other contexts (von der Gracht, Vennemann, and Darkow 2010). Ruff's (2015) practical reflections from the automotive industry have included early detection in new business environments and exploration and development in new business areas, among other things, as such key attributes. Sarpong and Maclean (2016) also argue that strategic foresight activities are often generally neglected and its practice would be improved if they took into greater consideration low-level employees, drawing their research also from the technology sector. As women have traditionally suffered barriers to excel to senior positions, any form of increased involvement in foresight activities at lower organisational levels would mean the need to recognise the views of women.

Recent literature has implicitly addressed the issue of impediment within foresight methodology for technology and innovation management. For example, Vecchiato and Roveda (2010) highlight the importance of social drivers of change in the environment, which include events in social, economic, political and ecological landscapes that surround an industry, but do not mention the rise of the importance of women in the new millennium. Battistella and De Toni (2011) also stress the importance of technological foresight requiring a different and specific treatment that relates to the strategic direction of the organisation for future orientation, but do not include implications. Similarly,

Rohrbeck and Gemuenden (2011) argue that corporate foresight that enhances innovation capacity must also serve a strategist role that explores new business fields as well as an initiator role that increases the number of innovation concepts and ideas. Perhaps this may be in the form of recognising 'new-wave innovation policies' that include corrective, disruptive and creative roles (Georghiou and Cassingena Harper 2011) that encourage breakthrough events and stimulate networks for future sustainability. Currently, only around 25% of IT positions are held by women (Peck 2015), and only 20% of chief information officer positions in Fortune 250 companies were held by women in 2012 (Khanna 2013). Hence, this beckons the need to recognise this shortage and its implications for corporate foresight activity in the technology sector.

3. Guiding hypotheses

As the research traces the origins of the current assumptions held about women in management and their ability to carry out tasks and hold decision making positions, the research explores classical management theories to question their applicability in the technology sector.

3.1. Shortage of women in technology

The gap between women and men holding senior positions is a marked recognition (O'Neil 2011). For many, the transition from middle to upper management positions is improbable (Eagly 2003), and those who do are the exception. One reason might be because the shortfall itself might lead to the lack of data on how women actually behave in senior positions, creating a vicious circle of there being few in the first place (Cole 2004).

A classic view of careers development is that known as the Peter Principle (Peter 1969), which stipulates that employees are appointed to their highest level of <u>in</u>competence (not competence). The shortage of information, and therefore reduced confidence in women, and fear that incompetence may have far more significant consequences in the technology sector, will imply that women will be appointed to lower levels – ie, to the position of competence, than incompetence. Similarly, it is thought that women are more task-oriented, rather than the traditional stereotypical views of gender roles. In this way, it is hypothesised that: *H1 – a woman's perception of the skills needed for a position she has acquired is greater or equal to that of her actual appointed role.*

Similarly, society requires women to prove themselves more in employment than their male counterparts (Barsh 2011), while they blame themselves for any failure (Adams and Kirchaier 2013). They underestimate their skills, abilities and performance, relative to masculine tasks (Beyer 1998), and take a greater hit on self-confidence when receiving negative feedback than their male counterparts (Roberts and Nolen-Hoeksema 1989). Public media has also been hostile, as when Facebook's COO, Sheryl Sandberg, went public *The New York Times* reported her appointment as being 'lucky' than suggested it as the outcome of her hard work, or in her view, 'having powerful mentors along the way' (Perlroth and Miller 2012).

Kanter's (1977) theory of Tokenism relates to a woman's negative experience during the job and inability to achieve equality being due to their token status – that is, their low proportion in a male-dominated workplace (Adams and Kirchaier 2013). It is likely that a woman's position in a male-dominated workplace will improve if their proportion is increased, thus improving their token status (Torchia 2011). This is more easily achieved if their exposure to what may be deemed 'the norm' from an earlier age to women in technology sectors is greater – for example, more role models, etc. Hence, it is hypothesised that: *H2 – there is a positive association between the availability of early exposure instruments to women and their subsequent proportion of presence in the technology sector of appointment.* The nature of work for women is evidently more difficult, and women need to overwork to overcompensate (Sandberg 2015). Some work places penalise mothers in professional jobs and affect progression (Glass 2004), providing evidence of a Glass Ceiling Effect.

3.2. Encouraging women into the technology sector

With the biases in mind, women by college stage are already anticipating the trade-offs between professional and personal goals (Lloyd 2010). This is worrying, as the academic performance of young girls at college is higher than that of boys, and in the US they earn 57% of the undergraduate and 60% if master's degrees (National Center for Education Statistics 2014); while in the UK, similarly, women make up 56% of undergraduate degrees (Higher Education Statistics Agency 2014). This implies that capabilities for achieving high exist, but something else stands in the way, possibly that of willpower, desire, or simply the lack of encouragement.

The early Expectancy Theory of Vroom (1964) stipulates that a person is motivated to select a specific behaviour over others due to the expected result of the selected behaviour, so the motivation is due to the desirability of the outcome (Lunenberg 2011). For the technology sector, women entering male-dominated industries may be psychologically hindered by anticipation of the unattainable outcome (Pearson 2012). Hence, it is hypothesised that: *H3 – there is a positive association between women's confidence in the technology sector and the seniority of the position held.* This is influenced by early exposure along the journey, and is based on a role congruity view (Eagly and Diekman 2005) that discourages take-up of senior positions due to prejudices of less-favourable behaviour by women in the workplace (Karen 2002). The greater women feel a lack of belongingness from early on, the greater likelihood this will result in reduced entry (Eagly 2003), and evidence suggests prejudice is highest where inconsistencies in gender roles already are manifest (Wang 2013). So it is also hypothesised that: *H4 – past experience of stereotypical behaviour negatively affects entry into male-dominated industries.*

3.3. Constitution of women

The most fundamental of this assumption is the biological make-up of a woman affecting the practical ability to take up a role (Shields 2013), and different expectations are associated with their gender (Li 2011). Eagly (2009) identifies a number of emotions relating to women, and the greater the 'emotionally expressive' the more negative impact she will have in the male-oriented workplace, such as in the technology sector. Hence, it is hypothesised that: *H5 – there is no association between the biological sex of a person and their attainment of the skills required for employment in the technology sector.*

The application of work is affected by social orientation (Barau 2012), and the most basic of all may be that relating to gender, most strongly affected by nurture, separate from that of biological sex. Differently, gender is the way culture defines and constrains the difference in terms of the male/female dichotomy, so gender and the labels associated with it (feminine and masculine) are neither fixed nor immutable (Siann 1994). Through social orientation, both genders grow traits and can be learned through situations (Schippers 2007). While there is not enough research on non-heterosexuals and their working environments (see Johnson and Repta 2012), the technology sector is still obviously one that nurtures creativity and genders can also be adaptive. So, we hypothesise *H6 – there is an association between the gender (non-biological sex) of a person and their attainment of the skills required for employment in the technology sector.*

4. Research methodology

4.1. Data collection

Data were collected by interviews and via an online questionnaire.

Interviews were employed to obtain in-depth understanding of the experiences women had at various levels of seniority in the technology sector. Purposive and snowball sampling were used, and initial access was successful because the researcher group included an active member of some feminist debate forums via the professional networking website, LinkedIn. It is acknowledged that interviewees solicited from specialist forums may have biased views relating to their forum topic, but this potential bias is balanced out by the need to obtain rich discursion that other less knowledgeable women not from such forums could not offer. The interviews with the five participants were face-to-face and lasted around an hour, and were digitally recorded and transcribed verbatim to ensure the richness was not lost. The interviewees requested anonymity and are summarised in Table 1.

Using both the themes identified from the interviews, and the extant literature, an online questionnaire obtained responses from targeted technology professional network platforms, both to triangulate in general and to ensure biases would be reduced. The questionnaire link was sent to around 700 invitees and 365 responded. The questionnaire was deliberately simple and based on measures from Bergman and Hallberg (2002), which contained all the questions within one scrollable page, requiring only about five minutes to complete. Measures were mainly in the form of 1–5 Likert scales.

4.2. Data analysis

Data analysis involved the following.

(i) *Development of themes:* Following Strauss and Corbin (1990), themes from the interviews were obtained through content analysis of the issues discussed. For this, three broad categories were established, using the simple codes of A, B and C respectively. In-vivo coding was used, resulting in a coding tree as given in Figure 1.

(ii) *Hypotheses testing:* A number of statistical tests were performed to test the specific hypotheses, loosely as a 'hook' for deeper re-examination of the interview material in part III of the research. Hence, given also the reasonably small sample of the data collected, the need for robustness testing was played down, and the results were reported for simplicity of interpretation.

(iii) *Case narratives:* Drawing on the results of the hypotheses in part II being either rejected or not rejected (supported), explanations for the same were sought from the interview transcriptions of part I. Broadly within the three themes of the study, supportive commentaries were selected based on dominant themes revealed in the in-vivo coding.

5. Findings

5.1. Inductive themes from interviewees

The dominant themes derived from the interviews were used to formulate the questionnaire, and presented in Table 2. The high frequency themes from the inductive interviews as shown in the coding tree above were extracted and formed the basis for the questions in the questionnaire for part II of the data analysis. These themes were augmented with the extant literature to complete the set of questions to measure specific variables as part of the hypothesis testing.

5.2. Hypotheses testing

The respondents were all in the technology sector, mainly female in biological sex (75.6%), heterosexual (89.4%) and not legally committed to a partnership (68.4%). The sample therefore adequately

Table 1. Interviewee profile.

Interviewee	Position held	Years in position	Industry
Ms V	Chief Operating Officer	1	Insurance
Ms W	Director	4	Information Technology and Services
Ms X	Director	5	Medical Devices
Ms Y	Global Head	5	Information Technology and Services
Ms Z	Chairwoman	5	Telecommunications

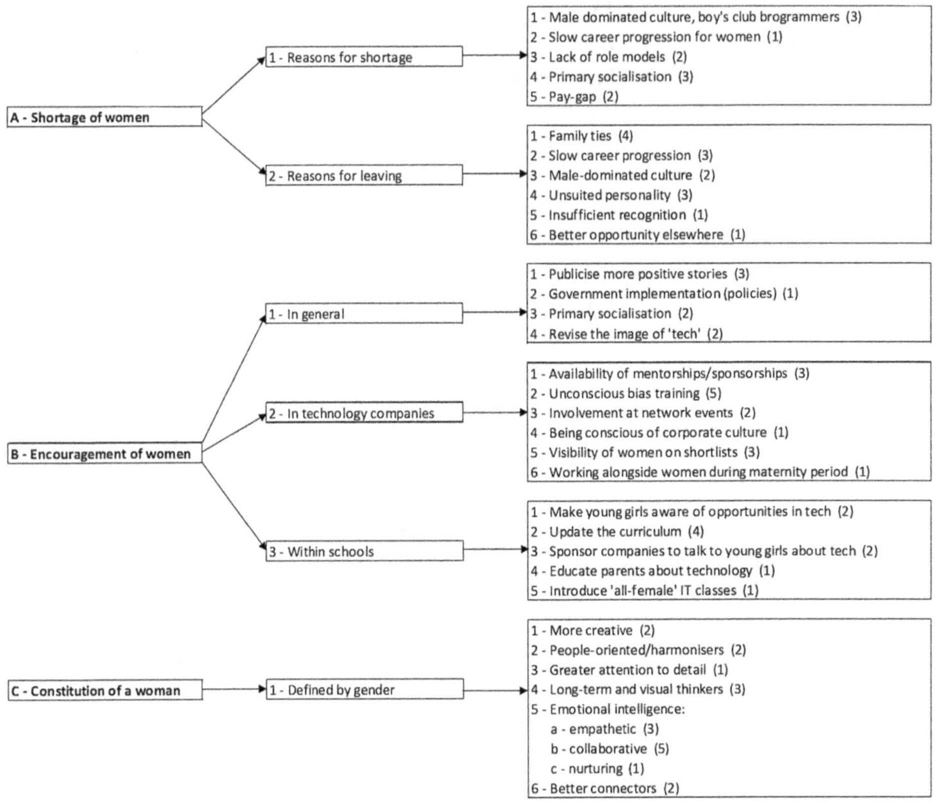

Figure 1. Themes from interviews.

represents the traditional population of detrimental women in technology as documented in the popular literature.

The study finds that the response of males and females are quite similar to their belief that they have achieved the skillset needed for their appointment (Male mean = 2.22; Female mean = 2.19), and the difference is statistically insignificant ($p = 0.771$). To test if the hypothesis (H1) stands that the Peter Principle does not apply to the technology sector, both biological sexes' skillset acquired need to be higher (lower in value) (mean = 2.21, SD = 0.871) than post of appointment held (mean = 3.81, SD = 1.203). In this case, the paired mean difference (−1.602) is statistically significant ($t = -25.002$, $df = 353$, $p = 0.000$), and therefore this part of the hypothesis is not rejected. To cross-check for its applicability for women only, the selected cases ($N = 269$) of women's belief of skills

Table 2. Variables for hypotheses testing.

What/Theme from interviews	Code from coding tree	Specific questions in questionnaire
Male-dominated culture	A.1.1. & A.2.3.	Q11
Lack of role models	A.1.3.	Q14a
Barrier to Glass Ceiling	A.1.6.	Q12
Family ties	A.2.1.	Q7, Q16a
Unsuited personality	A.2.4.	Q15
Publicise positive stories	B.1.1.	Q14
Availability of mentors/sponsors	B.2.1.	Q14
Unconscious bias training	B.2.2.	Q11
Involvement at networking events	B.2.3.	Q14b
Visibility of women on shortlists	B.2.5.	Q14c
Creativity	B.1.1.	Q17b
Emotional Intelligence	B.1.5	Q16c

acquired (mean = 2.2, SD = 0.908) were also greater (lower in value) than post of appointment held (mean = 3.78, SD = 1.194), with a paired mean difference (−1.58) that is statistically significant ($t = -21.775$, $df = 268$, $p = 0.000$). While we acknowledge the potential distortion in the use of self-reported perceptional data in tests relating to the Peter Principle, due to participants' inability to make accurate judgements, together, full support for H1 is nonetheless found.

There is weak support found for H2 that suggests a positive association between early exposure instruments of a woman and the extent of her future employment being technology-focused. Only the presence of role models (mean = 2.98, SD = 1.27) exhibited a statistically significant ($p < 0.05$), but reversed ($S\beta = 0.137$, $R^2 = 0.019$), association with the predictor variable. This could be because young women are deterred by the negative experiences and stories portrayed by role models, rather than used them as a positive motivator for entering employment in a more technology-focused sector. Mentors and network contacts do not indicate any predictive power. Hence, there is only partial support found for H2, which otherwise can be rejected entirely (model 3) if all variables are regressed together.

In encouraging women into the technology sector, support for H3 on the association between a woman's confidence and seniority of position held was also only partially found. The characteristic of being a risk-taker was the only statistically significant ($p < 0.10$) predictor of the seniority of post held ($S\beta = 0.126$, $R^2 = 0.03$). One reason may be that uncertainty of success in the sector is high and only taking risks is strong enough a motivator for achieving a high-level position. Ambition and boldness are weak predictors. Similarly, only weak partial support is found for H4 on the negative effect of experiencing stereotypical behaviour on entering a male-dominated industry. In this regression model it is clear that suppressor variables existed, which altered the strength of the predictor coefficients depending on the order of how the model was built. In this way, as can be seen in Table 3, the stereotype of child-bearing would be highly statistically significant ($S\beta = 0.22$, $p < 0.01$) without the other complementary predictors, but the predictive power ($R^2 = 0.048$) is much improved in the third model ($R^2 = 0.101$), where child-bearing is no longer a statistically significant predictor, but the stereotypes of biological sex ($S\beta = 0.173$, $p < 0.10$) and being overly emotional ($S\beta = 0.138$, $p < 0.10$) are. The reason why they cancel each other out might be because of autocorrelation, and such factors as child bearing are naturally related and specific to biological sex (Tables 4 and 5).

As a broad guide for examining the constitution of a woman, H5 and H6 respectively hypothesised the nominal association between biological sex and chosen gender on (the respondents' belief of) achieving the skills specific and necessary for working in the technology sector. H5 is almost fully

Table 3. Questionnaire data sample.

	N	%
Biological sex		
Male	85	24.4
Female	263	75.6
Chosen gender		
Masculine	91	26.1
Feminine	257	73.6
Sexuality		
Heterosexual	311	89.4
Non-heterosexual	37	10.6
Ethnic background		
White	271	77.9
Non-white	77	22.1
Name		
Anglicised	233	67.0
Non-Anglicised	115	33.0
Marital Status		
Married	110	31.6
Non-married	238	68.4
Have dependants		
Yes	97	27.9
No	251	72.1

Table 4. Test of statistical differences within groups of careers.

		Mean	*N*	SD					
					Paired tests of differences				
Both m/f	Skills set acquired	2.21	354	0.871	Mean	SD	*t*	*df*	*p*
	Post appointment	3.81	354	1.203	−1.602	1.205	−25.002	353	0.000
f only	Skills set acquired	2.2	269	0.908					
	Post appointment	3.78	269	1.194	−1.58	1.19	−21.775	268	0.000
					Independent tests of differences				
Men	Skills set acquired	2.22	85	0.746	Mean difference	*F*	*t*	*df*	*p*
Women	Skills set acquired	2.19	271	0.911	0.032	2.537	0.291	354	0.771

supported, except for creativity as skill which does have statistically significant association with biological sex ($\chi^2 = 3.01$, $df = 1$, $p = 0.083$), although the strength of the association is somewhat weak ($\varphi_c = 0.093$). This attribute is long believed to be sex-dependent so there might have been some bias in the self-reporting process. H6 however is rejected as the hypothesis that gender does have an association with the belief of achieving the various skills does not hold. The sample included both biological sexes as only a few women ($N = 6$) declared themselves of a different gender orientation from that of their biological sex to select cases to make meaningful statistical inferences (Table 6).

5.3. Case narratives

Three themed findings have emerged and are presented in this section as cases A, B and C, and their relationship is given in Figure 2. The figure shows that the constitution of women is the 'complex complex' (case C), denoted by a solid box that is difficult to penetrate, but composed of numerous (*X*1, *X*2 … *Xn* of) attributes, conceptualised as the 'perfect profile' (case B), all of which are either neutral when otherwise positive in non-technology sectors, called 'inversion', or suffering serious consequences, called 'contraction' (case A).

Table 5. Regression results.

					Sβ for Models		
		Mean	*N*	SD	1	2	3
H2	*Extent tech sector*	1.99	350	1.124			
	Role models	2.98	350	1.270	−0.137**	−0.063	−0.061
	Mentors	3.15	350	1.250		−0.098	−0.095
	Network contacts	2.79	350	1.236			−0.009
	Constant				2.346	2.429	2.437
	F				6.627	1.483	0.019
	R-square				0.019	0.023	0.023
H3	*Post appointment*	3.81	350	1.215			
	Risk-taker	2.55	350	1.022	0.142**	0.177**	0.126*
	Motivated by ambition	1.92	350	0.953		−0.075	−0.095
	Bold	2.32	350	0.984			0.098
	Constant				3.383	3.462	3.384
	F				7.191	1.560	2.031
	R-square				0.020	0.025	0.030
H4	*Extent male-dom industry*	2.24	354	1.117			
	Child-bearing stereotype	3.10	354	1.470	0.220**	0.083	0.050
	Biological sex stereotype	3.53	354	1.402		0.250**	0.173*
	Overly emotional stereotype	3.36	354	1.449			0.138*
	Constant				1.722	1.343	1.279
	F				17.860	16.915	3.600
	R-square				0.048	0.092	0.101

p values, $* = p < 0.10$, $** = p < 0.05$.

Table 6. Associations of nominal tests.

		χ^2 value	*df*	*p*	φ_c value
H5	Sex –> Skill, innovative	0.079	1	0.779	0.015
	Sex –> Skill, creative	3.010*	1	0.083	0.093
	Sex –> Skill, problem solve	0.191	1	0.662	0.024
	Sex –> Skill, fast thinking	1.908	1	0.167	0.074
	Sex –> Skill, adaptability/flex	0.012	1	0.914	0.006
	Sex –> Skill, customer focused	0.258	1	0.612	0.027
H6	Gender –> Skill, innovative	0.346	1	0.556	0.032
	Gender –> Skill, creative	0.980	1	0.322	0.053
	Gender –> Skill, problem solve	0.014	1	0.905	0.006
	Gender –> Skill, fast thinking	1.500	1	0.221	0.066
	Gender –> Skill, adaptability/flex	0.003	1	0.954	0.003
	Gender –> Skill, customer focused	0.046	1	0.830	0.012

p values, * = $p < 0.10$.

5.4. Case A: inversion and contraction

The negativity of role models identified in H2, which seems not to encourage direct entry into the technology sector, but has the opposite effect, and the extent to which childbearing seems a normalised life event for women, can be characterised by the overall phenomenon of 'inversion and contraction'. This means, there is an inversed expectation that 'contracts' the position of women in the sector (notwithstanding the pun of contraction in childbearing!). These biases, not hitherto known, can seriously affect business development in the immediate term and corporate foresight activity in the longer-term future. As an interviewee put it simply: '*when trying to get these women into more senior positions it was almost like their careers had stopped*' [Ms Z].

'Inversion' is when something ordinarily positive seems not to have a positive outcome for the women concerned. The consequences are not necessarily negative, but are neutral when they should under other normal circumstances be deemed celebratory and promotional for the person concerned. Such things should ordinarily improve a person's professional vitae (resumé) and considered key in building a professional career. An example of being well-networked was commented as follows:

> I think that once you hit a particular level in technology, roles aren't advertised to you as much .... That's why it's important to have a good network ... I think that's the biggest challenge I had faced. I didn't build my network early enough. It's a secret job market. [Ms Y]

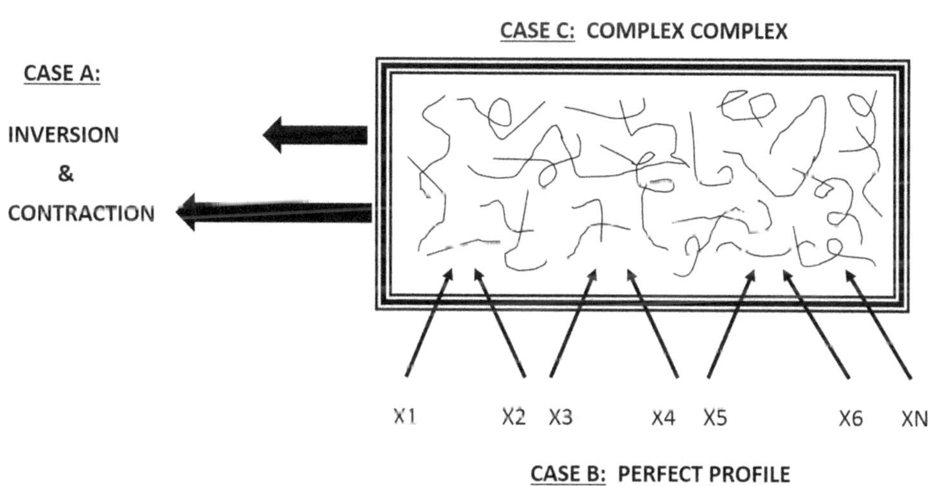

Figure 2 . Relationship of case narratives.

In this situation, Ms Y's attributes had gone beyond internal organisational normal working and received external recognition in being established as part of other professional networks. However, it seems this advantage was not enough to bring forth any benefits for her, and she believed this needed to have materialised even earlier and received greater prominence to be infallible so that others in the workforce could not attack it. The secret job market Ms Y referred to is evidence that the benefit is ignored, and hence illustrates the 'inversion' effect.

Examples of other inversion effects could include having good work ethics, possessing specific qualifications for a wide range of skills, additional vocational activities, or life events like having grown-up children who themselves are well-networked and belong to learned associations. Instead of reaping in-company benefits, these seemed as threats to the male workforce and were therefore brushed off, or subtly ignored. As the judgement of women on the above is most likely different from that of men, the inversion effect is not just a threat to the specific women concern but also to womankind in general if their unique take on corporate foresight is also ignored.

Similar to inversion, but a much more severe form, 'contraction' takes place when the established or normal positivity actually makes the situation worse than before. In the case of Ms V, given as the example below of starting a family, it would seem better not letting others know about it, so that inversion is at least better than contraction:

> I just said that it sucked at the thought of having to have a family one day with having my business ... and [a colleague] completely used that against me. He bought all my employees away from me at double the wage and said that he did that because I am ... a woman. I've had many challenges about being a woman but that was my worst ... It taught me a lot about how men can twist things ... [Ms V]

Had Ms V kept her family planning matters confidential, she would not have suffered the loss of employees that she was in charge of at the time. It is unclear why she was treated in the way she explained as the actions of the man involved seemed completely unrelated to the situation.

While the example given of Ms V might be extreme and singled-out behaviour, contraction is likely in practice to take place more subtly, so that any unfair demotion or detriment goes un-noticed. The trigger points can include a range of life events, skills and attributes like being young and energetic, and even good looks that threaten some men. The detriment could be in the form of reallocation of work responsibilities for more inferior tasks, and a degrading of social status. These actions might at first seem harmless, or at worst rude, but their consequences are long-lasting, if vital market and strategic foresight thinking are lost, which potentially could be important for shaping the future development of the organisation.

Inversion and contraction can take place independently, or together. They can happen because of the nature of the social phenomenon, or through the intervention of another actor, mostly that of a male-dominated influence. Other example commentaries of either of these incidences are expanded in Appendix 1.

5.5. Case B: perfect profile?

Among an array of findings, H1 and H3 seem to characterise the perfect profile of women suited for the technology sector. The name given to this conceptualisation is imperative, as it highlights the importance of 'perfection' and how a simple set of attributes does not suffice for women working in the technology sector; they must possess a *substantial* set of attributes, and exhibit them well for valuable and recognisable utility. The suggestion that the Peter Principle does not apply to the sector from the hypotheses testing infers a quality of women who are appointed to their worth, and is a risk-taker to have the confidence to work within its difficult environment. Among these wide and general descriptions of women, numerous adjectives were commonly used to express concerns about women in the workplace. The following couple of comments reflect the nature of female behaviour.

> [Women] bring a different aspect to the business. They have creativity, different opinions, they bring about a people-approach ... are more long-term orientated and are prone to questioning more. They offer a challenge

> … have a lot more attention-to-detail … and question whether things are right. They are good harmonisers and so people-focused … are visual thinkers. It's empathy as well. [Ms X]

> Women tend to create environments that are less competitive and more cooperative and collaborative. They tend to be more nurturing … more concerned with making sure everyone is moving along in their careers and are happily progressing … [S]ome have their traits and some do not. [Ms Z]

The narrative of Ms X focuses on qualities that are domain expanding, which are particularly valuable in external environmental scanning in corporate foresight exercises. In the psychology literature, they are associated with right-hemisphere brain activity or typical left-handedness. Our data did not confirm this, as this is not central to the research. However, these are attributes that are few in the general workforce, and are normally associated with historical figures of common knowledge, and less commonly recognised within the male-dominated workforce. Similarly, Ms Z's comment reflects the soft-systems nature of managing human resources. Cooperation, collaboration and humility are team-based skills that are central, not only to ensure good relationships between the male and female members, but also for creating good working environments free from friction and for brainstorming of innovative ideas; in essence, this is the purpose of corporate foresight exercises. These qualities are also empathetic in nature, which ultimately result in better customer service quality and understanding, which again are important antecedents of successful corporate foresight activity.

Indeed, while a few interviewee comments cannot establish an exhaustive list of the ultimate candidacy for appointment within the technology sector, there does seem to be negative consequences for a woman who does not behold 'positive' attributes. This is similar to case A of inversion and contraction, but here the focus is on the various attributes, and hence we argue the quintessential need to possess them as being neutral and not a negative impact per se. In other words, an individual's competitive advantage is not gained if a woman possesses these attributes in the technology sector, and equally their absence does not lead to a loss in individual competitive advantage; however, their unlikely absence might stand out and is a wasted opportunity for an organisation.

Ms X comments:

> I had an appraisal and I did really well for that year … [and] I genuinely thought I was going to get a promotion … [but] I didn't. The reason was they thought I was 'overweight' and I didn't 'look the part' … I have the highest performing sales in the company and have maintained that reputation every year, but the excuse for me not getting promoted all that time was because I apparently didn't have any gravitas. [Ms X]

Her complaint here focuses on three negative characteristics – being 'overweight', not 'looking the part' and not having 'gravitas'. These characteristics are subjective and often insignificant under ordinary organisational working conditions. However, it seems for women in the technology sector, any characteristics that are (even subjectively) deemed negative would have a detrimental effect on the women concerned, which in other sectors might be treated neutrally.

Other examples of what might have constituted the 'gravitas' are given in Appendix 1. These include: attention to detail, long-term orientation, inquisition, good harmonisers, and people-focused. The restrictive characteristics, indicating frivolity, include: holding back in situations, lacking of confidence, non-self-promoting, and less-competitive. Where women try to take up the restrictive behaviour, they would be seen the negative side of it. As Ms Z describes, *'I have experienced being called "bossy" rather than being called a "leader"* … *[and] even called "mean" by another member on the board'*. Hence, there are strong and prominent sets of attributes that are apparent, some which are naturally positive but ignored, and some which are neutral or subjectively perceived negative that have detrimental effects. The collective of these attributes that sustain women in technology management thus establishes the case of 'perfect profile'.

5.5. *Case C: the complex complex*

In trying to understand what constitutes a woman, which is perhaps the root cause of the shortage problem, it is found that biological sex and gender are only surface level issues. Indeed, there may be more detailed biological and medical explanations for the minutiae of this genetic constitution, but from the management perspective, H5 and H6 identify the woman herself is a complex (*noun*: a closed system) which itself is extremely complex (*adjective*: difficult to comprehend extensively), hence being termed a 'complex complex' for women in technology. It (she) is therefore a difficult-to-penetrate closed system that is inherently rich in advice offering. Complexity is used to attribute the complex not simply because issues are too complicated (by this, an intelligent person can still learn the technicalities of the issues) but because they are open to numerous interpretations and advance assimilation, and thus requires a greater degree of empathy.

Interviewees struggled to comment directly on this issue. However, they did talk about women-specific issues, such as menstruation and child-bearing, which are difficult for men to comprehend or empathise with extensively. The reason why these issues constitute part of the complex complex is because, while there are medical experts on these matters, their study is comparatively imprecise as it is based heavily on the psychological aspect as much as the physical composition of the human being. Other examples of the complex *noun* might include their debatable multi-tasking ability and high EQ psyche. Similarly, examples of the complex *adjective* might include: ambiguity of body language and interpretation of tones of conversation, differences in choice of vocabulary, and interaction differences, among potentially numerous others.

Sometimes it is not easy to see why women do things, choose to enter a sector or offer up explanations for their approach. In that sense, the basis for that complexity is unexplained altogether. For example, Ms W comments:

> I'm not a techy person at all and I've recognised that there are more men in the engineering, programming side of tech, but if I wanted to do something like that, I wouldn't let anyone stop me ... It's based on your own personal skills, abilities, actions and behaviours. [Ms W]

The above quotation suggests that decisions might be made by impulse, rather than through logical reasoning, but the determination in following through the decision ensures its completion. The value though is that while these impulses are inherently built upon an internal structure that defies logic, they do seem to work, and the ultimate action is socially accepted.

Hence, the complex complex is a phenomenon that is easier to admit its existence than to define its operation. However, its implications for corporate foresight within the technology sector are significant if the complexity is what discourages the inclusion of women in foresight activities. The diversity of interpretation that makes the complex particularly complex is a potential loss if not capitalised, as foresight activities are best operationalised through unusual and multiple dimensions, which gives it a valuable source of competitive advantage for the organisation if included.

6. Insights for managing corporate foresight

The above findings offer insightful suggestions for ex-ante corporate foresight activity, and hence also for corporate foresight methodology, specifically within the technology sector. The primary research of this paper did not set out to investigate how exactly such methodology requires changing. However, both the quantitative and qualitative aspects have identified distinctive characteristics about why women are few in the technology sector but are of potential value and could be included more in corporate foresight exercises to reap the benefits they discussed. These suggestions are summarised in Table 7, and discussed as follows.

6.1. Environment, orientation and recruitment

Starting with the extant understanding on corporate foresight and calls for new innovations in methodology (eg. Georghiou and Cassingena Harper 2011), methodology must move away from a standard external environmental consideration of political, economical, social, technological and legal factors (ie. a standard PESTEL analysis, presented herein as PEcSTEL to differentiate economical

Table 7. Comparison of existing methodology and new implications from the research.

Dimension	Existing Foresight	New Implications	Supporting Evidence
Environment	PESTEL	PEcSTEL + W+w	Generally
Orientation	Future	Future, with anticipation of increased female markets	Generally, & H4
Engagement in foresight	Normal	Increased effort to recruit women in foresight activity	H1, H2
Division of foresight work	All undertake equal foresight activity	Use women (who are mainly high risk-takers) to engage in diversification heavy scenarios	H3
Cognitive reasoning	Relate to present and normal past	Relate deeply to women's past personal childhood	H4
Nature of industry	Involve random panel for foresight	Involve panel of more women if involving creative industries	H5
Decision making process	Perceived equal assessment of all employee opinions	Recognise small skills more to overcome an inherent bias	Case A
Recruitment panel	Involve sex proportions as recruited	Involve women dominant panels in foresight activities involving 'controversial' cases	Case B
Conclusion drawing process	Expect detailed derivation of solutions at all stages	Allow 'leap' in strategizing decision outcomes	Case C
Example	Scenario planning	STEM specific foresight	All

from environmental factors) to one that more explicitly recognises the addition of women (W), where it affects the total of PEcSTEL, plus a smaller constant influence (w). In this way, foresight methodology, F, for the technology context ($_{\text{TECH}}$) is:

$$F_{\text{TECH}} = f(P, Ec, S, T, E, L, W, w)$$

$$F_{\text{TECH}} = W[\alpha + b_1P + b_2Ec + b_3S + b_4T + b_5E + b_6L + \varepsilon] + w$$

The present research indicates this importance of this category and cannot regard it as a variable within either the social or narrow ordinary technological factor (eg. not: $P, Ec, S[+W_1], T[+W_2], E, L$). In the same way, the orientation of foresight methodology may benefit from incorporating the default prominent anticipation of a growing female future market.

This suggests the need to engage in corporate foresight recruitment panels to involve a disproportion in favour of women in particularly 'controversial cases' where women are found to possess a stronger capability to exhibit attributes of co-operation, harmony and empathy (this is evidenced by Case B from the research). This need is extended to concern if the nature of the industry is itself female-dominated or general, and if so, the panel of predominantly women are best serviced when concerning creative industries (eg. arts and media related technologies), as creativity is also is a predictive attribute from test of H5. The action to do so is not an isolated effort the few leading technology companies that compete for advantage and performance improvement should make, but one concerning global pressure for national governmental bodies to correct stereotypes about women in various careers, particularly STEM ones, at early stages to ensure the unique attributes women possess that are invaluable in sectors requiring innovation can be benefitted from them in both normal strategic, as well as corporate foresight, activities for the sustainable long term future. The use of women and focus on their (perhaps biased) views during market scanning exercises as part of corporate foresight activity have the specific advantage of utilising a much larger, fast growing and more easily identifiable market of female-related products than other demographic groups (eg. sexual, ethnic and cultural minorities).

6.2. Decision making processes in foresight methodology

The research did identify a difference of expectations of men and women in the technology sector (based on H1) and that women are appointed more to the skillset they possess, despite some potential bias in the sample of respondents surveyed. Entry into STEM sectors may have been affected by the stereotypes experienced in women's early life (based on H2), so the technology sector does have distinctive characteristics, for which some specialist skillsets, such as creativity and risk-taking (based on H3), can benefit. The suggestions for foresight methodology concern how women are best utilised for decision making and drawing of specific conclusions, as they concern mainly the cognitive reasoning, when otherwise normally relates to one's present and normal past, but for women their past relates more strongly to a personal childhood (H4) which flavours their ultimate decisions. The qualitative research (Case C) also supports this view that while a 'complex' characterises women, it is beneficial to allow 'leaps' in the conclusion drawing process; in this way, it is not always necessary to require clearly articulated derivations for solutions, which might only add further restriction to, and potential compromise of, pushing out frontiers in innovation for the industry whole.

A suggestion is that technology firm corporate foresight strategists should encourage greater inclusion of women in general, and to identify women with more creativity and boldness, in such exercises. The division of foresight work should be separated, so that business and innovation related decisions concerning highly diversified scenarios are undertaken predominantly by women, who are high risk-takers (H3). While this approach might be seen as positive discrimination and working against equality principles, it is likely to be neutralised by the long-term inherent biases against women (Case A), and calls are placed on the further need to recognise small skills that might otherwise be a lost oversight.

7. Conclusion

We end by highlighting the value and promise of the research data, and reiterate the importance of practice to recognise our findings. Our overall research approach was deliberately simple: our guiding hypotheses provided a gauge of the position of the shortage of women in technology, while the case interrogations examined more explicitly the underlying issues that can potentially improve corporate foresight activity. While our data were few in quantity and limited in some self-reporting perceptions, they were nonetheless rich in purposive value. The narrowly targeted sample ensured high respondent participation, but we are bereft of the complex biological, psychological, or even technical differential minutiae of women in management research findings, which are the research's limitation. Without embarrassment or apology, we believe the simple yet strong messages of narration are an important contribution to the extant literature: to make explicit the inclusion of women in corporate foresight activities within the technology sector in the light of the suggestions made. How exactly to use women within corporate foresight exercises and established methodologies is however the subject of further and future research within the field of technology analysis and strategic management!

Disclosure statement

No potential conflict of interest was reported by the authors.

References

Adams, B. R. and T. Kirchaier 2013. "Making it to the Top: From Female Labour Force Participations to Boardroom Gender Diversity". Financial Markets Group Discussion Paper 715, London School of Economics, ISSN 0956-8549-715.

Barau, A. B. 2012. "Gendering the Digital Body: Women and Computers." *AI & Society* 27: 465–477.

Barsh, J. L. 2011. *Special Report: Unlocking the Full Potential of Women in the U.S. Economy*. McKinsey & Company. Accessed December 15, 2015, http://www.mckinsey.com/Client_Service/Organization/Latest_thinking/Unlocking_the_full_potential.aspx.

Bergman, B., and L. R. -M. Hallberg. 2002. "Women in a Male-Dominated Industry: Factor Analysis of a Women Workplace Culture Questionnaire Based on a Grounded Theory Model." *Journal of Sex Roles* 46: 311–322.

Battistella, C., and A. F. De Toni. 2011. "A Methodology of Technological Foresight: A Proposal and Field Study." *Technological Forecasting and Social Change* 78: 1029–1048.

Beyer, S. 1998. "Gender Differences in Casual Attributions by College Student of Performance on Course Examinations." *Current Psychology* 17 (4): 346–347.

Cole, N. D. 2004. "Gender Difference in Perceived Disciplinary Fairness." *Gender, Work and Organization*, 11 (3): 254–279.

Eagly, H. A. 2003. "Finding Gender Advantage and Disadvantage: Systematic Research Integration is the Solution." *The Leadership Quarterly*, 14, 851–859.

Eagly, A. H. 2009. "The His and Hers of Prosocial Behaviour: An Examination of the Social Psychology of Gender." *American Psychologist* 64 (8): 644–658.

Eagly, A. H., and A. B. Diekman. 2005. "What is the Problem? Prejudice as an Attitude-in-Context." In *On the Nature of Prejudice: Fifty Years After Allport*, edited by Glick Dovidio and L. Rudman, 19–35. Malden: Blackwell.

FT. 2015. "Gender Equality in the Tech Sector Will Benefit the Global Economy." *Financial Times Newspaper* [Online], Accessed July 2, 2016, http://www.ft.com/cms/s/0/e2f8ad0a-bdd6-11e5-9fdb-87b8d15baec2.html#axzz3xgWhH6MG.

Georghiou, L., and J. Cassingena Harper. 2011. "From Priority Setting to Articulation of Demand: Foresight for Research and Innovation Policy and Strategy." *Futures* 43 (3): 243–251.

Glass, J. 2004. "Blessing or Curse? Work-Family Policies and Mother's Wage Growth Over Time." *Work and Occupations* 31 (3): 367–394.

Higher Education Statistics Agency. 2014. "Table 13: HE Qualifications Obtained by Sex, Subject Area, Level of Qualification Obtained and Class of First Degree 2009/2010 to 2013/2014." Accessed January 9, 2016, http://www.hesa.ac.uk/stats/.

Johnson, J. L., and R. Repta. 2012. "Sex and Gender: Beyond the Binaries." In *Desiging and Conducting Gender, Sex and Health Research*, edited by J. L. Oliffe and L. Greaves, 17–38. London: Sage.

Kanter, R. M. 1977. *Men and Women of the Corporation*. 2nd ed. New York: Basic Books.

Karen, A. H. 2002. "Role Congruity Theory of Prejudice Towards Female Leaders." *Psychological Review* 109 (3): 573–598.

Khanna, D. 2013. "We Need More Women in Tech: The Data Prove It." *The Atlantic*, October 29, http://www.theatlantic.com/technology/archive/2013/10/we-need-more-women-in-tech-the-data-prove-it/280964/.

Lewis, T. 2014. "Apple's Health App Tracks Almosts Everything … Except Periods." [Online], Accessed July 2, 2016, http://www.livescience.com/48040-apple-healthkit-lacks-period-tracker.html.

Li, D. B. 2011. "Sex Differences in Cooperation: A Meta-Analytic Review on Social Dilemmas." *Psychological Bulletin* 137 (6): 881–909.

Lloyd, D. L. 2010. "Expertise in Your Midst: How Congruence between Status and Speech Style Affcts Reactions to Unique Knowledge." *Group Processes & Intergroup Relations* 13 (3): 379–395.

Lunenberg, F. C. 2011. "Expectancy Theory of Motivation: Motivating by Altering Expectations." *Internatioanl Journal of Management, Business and Administration* 15 (1): 1–5.

Massari Coelho, G., A. C. Filgueira Galvao, A. C. Guedes, I. A. Carneiro, C. Nehme Chauke, and L. Fellows Filho. 2012. "Strategic Foresight Applied to the Management Plan of an Innovation Development Agency." *Technology Analysis and Strategic Management* 24 (3): 267–283.

National Center for Education Statistics. 2014. "Table 283: Degrees Conferred by Degree-Granting Postsecondary Institutions, by Level of Degree and Sex of Student." *Digest of Education Statistics*, http://nces.ed.gov/programs/digest/d13/tables/dt13__318.io.asp.

O'Neil, D. H. 2011. "Do Women's Networks Help Advance Women's Careers? Differences in Perceptions of Female Workers and Top Leadership." *Career Development International* 16 (7): 733–754.

Pearson, S. Y. 2012. "Organizational Citizenship Behaviour in IT Professionals: An Expectancy Theory Approach." *Management Rsearch Review* 25 (12): 1170–1186.

Peck, E. 2015. "The Stats on Women in Tech are Actually Getting Worse." *Huffington Post*, http://www.huffingtonpost.com/2015/03/27/women-in-tech_n_6955940.html.

Perlroth, N., and C. C. Miller. 2012. "The $1.6 Billion Woman, Staying on Message." *Business Day*, February 4.

Peter, L. J. 1969. "The Peter Principle: An Explanation of Occupational Incompetence." *Management Review* 58 (2): 1–12.

Roberts, T.-A., and S. Nolen-Hoeksema. 1989. "Sex Differences in Reactions to Evaluative Feedback." *Sex Roles* 21: 725–747.

Rohrbeck, R., C. Battistella, and E. Huizingh. 2015. "Corporate Foresight: An Emerging Field with a Rich Tradition." *Technological Forecasting and Social Change* 101: 1–9.

Rohrbeck, R., and H. G. Gemuenden. 2011. "Corporate Foresight: Its Three Roles In Enhancing the Innovation Capacity of a Firm." *Technological Forecasting and Social Change* 78: 231–243.

Ruff, F. 2015. "The Advanced Role of Corporate Foresight in Innovation and Strategic Management — Reflections on Practical Experiences From the Automotive Industry." *Technological Forecasting and Social Change* 101: 37–48.

Sandberg, S. 2015. *Lean In: Women, Work, and the Will to Lead*. London: WH Allen.

Sarpong, D., and M. Maclean. 2016. "Cultivating Strategic Foresight in Practise: A Relational Perspective." *Journal of Business Research* 69: 2812–2820.

Schippers, M. 2007. "Recovering the Feminine Other: Masculinity, Femininity and Gender Hegemony." *Theory and Society* 36 (1): 85–102.

Shields, S. A. 2013. "Gender and Emotion: What we Think we Know, What we Need to Know and Why it Matters." *Psychology of Women Quarterly* 37 (4): 423–435.

Siann, G. 1994. *Gender, Sex and Sexuality: Contemporary Psychological Perspectives*. Oxon: Taylor and Francis.

STEM Careers. 2015. "Why STEM careers, Rolls-Royce." http://www.wherestemcantakeyou.co.uk/docs/Why_STEM_Careers.pdf.

Strauss, A., and J. Corbin. 1990. *Basics of Qualitative Research*. Newbury Park, CA: Sage.

Torchia, M. C. 2011. "Women Directors on Corporate Boards: From Tokenism to Critical Mass." *Journal of Business Ethics*, 102, 299–317.

van der Duin, P. A., and E. den Hartigh. 2009. "Keeping the Balance: Exploring the Link of Futures Research with Innovation and Strategy Processes." *Technology Analysis & Strategic Management* 21 (3): 333–351.

Vecchiato, R., and C. Roveda. 2010. "Strategic Foresight in Corporate Organizations: Handling the Effect and Response Uncertainty of Technology and Social Drivers of Change." *Technological Forecasting and Social Change* 77: 1527–1539.

von der Gracht, H. A., C. R. Vennemann, and I. Darkow. 2010. "Corporate Foresight and Innovation Management: A Portfolio-Approach in Evaluating Organizational Development." *Futures* 42 (4): 380–393.

Vroom, V. H. 1964. *Work and Motivation*. San Francisco, CA: Jossey-Bass.

Wang, A.-C. 2013. "Gender Makes the Difference: The Moderating Role of Leader Gender on the Relationship between Leadership Styles and Subordinate Performance." *Organizational Behavior and Human Decision Processes* 122: 101–113.

Wise Campaign. 2015. "Women in the UK STEM Workforce." Accessed September 7, 2015, https://www.wisecampaign.org.uk/resources/2015/09/women-in-the-stem-workforce

Appendix 1. Additional commentary in support of case narratives.

	Case A	Case B	Case C
Phenomenon	*Inversion and Contraction*	*Perfect Profile?*	*The Complex Complex*
Relate to Hs:	H2, H4	H1, H3	H5, H6
Narratives:	I think that once you hit a particular level in technology, roles aren't advertised to you as much can only really get new roles if you wish to, through your networks. That's why it's important to have a good network. If you have those work relationships with the right people, and then those people actually put in a good word for you, you'd get straight onto a shortlist. Also, the amount of people they are allowed to submit to the shortlist is limited, so they don't really want to put anyone on it that may cause risk. And I think that's the biggest challenge I had faced. I didn't build my network early enough. It's a secret job market. – **Ms Y** The career path for women could also put them off. In technology, you've got to be on top of the technology 24/7; it's very fast paced. I think that when women leave the organisation for a year and have children, they miss out a lot on what has been happening with the company. Technology changes within the blink of an eye and it is hard to keep up. By the time they come back, they feel out-dated	An example once in my career, when I had a male boss and I had an appraisal and I did really well for that year, I met all my targets and my colleagues were even so pleased with my performance that year, I genuinely thought I was going to get a promotion. Basically I didn't so I raised it. I needed an explanation, at least so I could improve. The reason was they thought I was 'overweight' and I didn't 'look the part' … I have the highest performing sales in the company and have maintained that reputation every year, but the excuse for me not getting promoted all that time was because I apparently didn't have any gravitas … – **Ms X** They bring a different aspect to the business. They have creativity, different opinions, they bring about a people-approach, and I just think these things can add strength and value to tech … Women are more long-term orientated and are prone to questioning more. They offer a challenge. I think this helps in delivering a more sustainable piece of technology. Women also	I'm not a techy person at all and I've recognised that there are more men in the engineering, programming side of tech but if I wanted to do something like that I would. I wouldn't let anyone else stop me … Q14: I think it's based on your own personal skills, abilities, actions and behaviours. – **Ms W** Without having a woman's opinion implemented into things, you won't be able to understand the full picture and provide solutions to it. If you apply this to the creation of apps, being women specific, men wouldn't alone be able to create an app directed for women because they don't know everything about a woman – only a woman does. Such as health apps, they offer pools of information but none of them monitor menstruation. A man would never think to monitor menstruation. If women aren't involved in some of these tech solutions that are being changed – one change in tech can change almost every industry – it's that significant! If women aren't involved technology won't be able to provide solutions to everything provided. If

(*Continued*)

Continued.

	Case A	Case B	Case C
	with the technology and perhaps lose their confidence and get intimidated by their colleagues when they return. Even taking a career break is never seen as a positive in this industry. The technology sector does not like part-time working. It's like a vicious cycle because the industry isn't helping itself move forward and recruit these women in because they are instantly put off of the cultures that have already been set up. – **Ms X** We do need more positive stories, more role models to look up to, and some tech companies are working on that. Tech City UK for example, has a group of 100 people that they profile every year and make sure they have a balance in women. So there are definitely some companies that are working on making sure everyone knows that there are some women working on interesting things. We need to also make sure young girls aren't getting the wrong idea about the sector. And we need to deal with people who have the bad attitudes. – **Ms Z** When I interviewed for this position a while back, they knew I was in my early 30s and information like this, and I just felt like I was being fished for information. Being a part of the interview process many times, I am familiar with certain behaviours and, basically I was explicitly asked: 'Are you going to have a child soon?' If so, basically own up now because we don't want you. I've been in companies where males in the interview panel have come out after interviewing women and been like 'Hmm … this woman is in her late twenties though … What if she has a kid?' I'm at a certain age now where I feel people do look at your differently. If a man was to go to an interview, I doubt he'd ever be asked, 'But, have you got any children?' Both parents care for their children so why is it that one parent faces prejudice!? I just feel that in technology, there is definitely a 'Boy's Club' mentality. My career would have been different if I was a man. I think I would have got better opportunities. I should have had director a year ago. – **Ms X**	have a lot more attention to detail I have found, and question whether things are right. They are good harmonisers and where they are so people-focused; they look at the person at the end of the technology. They are visual thinkers. It's the empathy aspect as well. – **Ms X** I do find that there are women like me, who give it everything and then are rewarded with a C-Level position. – **Ms V** I joined because a woman was the head of the company and, the board she put together was very balanced as far as gender went. I thought to myself 'the door is open here. Women can make it to the top.' And I certainly felt that. I liked the way that she ran the business and how collaborative the place was. She created a great corporate culture. However, she sadly retired and a man took over her. The culture changed entirely … The message to me was, even though I was a senior woman, you're not going to get anywhere here and the door is shut now. – **Ms V** Women do tend to hold back and not push themselves forward and this just makes things worse. We also need to teach women to be confident, push themselves forward, take risks etc. … . Q9: We don't like fuss, we don't like to promote ourselves, we don't boast about what we are doing. Whereas, generally speaking men are easier at doing that. – **Ms Z** The shortage of women in tech spurred me on. There have been some cases where I have been at tech events and I have been the only woman, but that makes me memorable. There weren't any other women speaking at conferences but that was a good thing for me personally, but not a good thing for women in general. I've worked with women who have tried to use their sexuality to get in rather than their brains and I think I actually tried to help a couple not to do that. I've just been focused on doing my job and doing it well. – **Ms Z** I have experienced being called 'bossy' rather than being called a 'leader'.	you look at things from just one perspective, you get a very narrow solution. – **Ms Y** Their collaboration. Their willingness to not be so directive and, create an atmosphere where both genders feel safe and comfortable in putting out new ideas. I feel that this would bring more innovation to tech companies. They will solve problems quicker. Women have a knack for helping people collaborate in creating a trusting atmosphere. You feel an atmosphere change in the room, feeling women around. There is more of a balance. Men can't act like blokes and it just opens up new avenues of discussion. – **Ms V**

(*Continued*)

Continued.

Case A	Case B	Case C
I made my own business right, so you can't be any higher up the food chain in your own business. There was an Irish guy who came to Prague – I had been in my business for about 6 years and was ready to leave. I never intended to start this business! It was interesting and fun but wasn't going to be my life. Anyway, this guy – because I was young, naïve, trusting and knew everyone in Prague. It is a very small town. I trusted people. He comes along and says, 'Oh, I'd really like to hear a lot more about your business. I'm interested in buying it.' – I should have gone the whole legal route and made him sign paperwork, but I didn't. I just went and had a drink with him, told him everything, told him too much. Within a week he went to all of my employees and told them all that I was a woman and that I wanted to start a family – I didn't tell him that, I just said that it sucked at the thought of having to have a family one day with having my business. And he completely used that against me. He bought all my employees away from me at double the wage and said that he did that because I am an American and I am a woman. I've had many challenges about being a woman but that was my worst … It taught me a lot about how men can twist things, not just men, but turning this whole thing into 'Oh, she just wants to have babies more than she wants to have a business … ' is like- ugh! Who says I couldn't have both!? … No other females had their own business in Prague except a few other American women or European women. Working in the UK corporate world – men would call it 'unconscious bias' but I wouldn't say it was that unconscious – some of the comments made. – **Ms V** The stereotypes have probably, at times, got in the way and tried to hinder my progress but that just made me want to prove them wrong. There have been times where I wouldn't have even be given eye contact, I've missed out on networking opportunities, there are a number of things. That's why I left my previous company and started my own. – Ms W	Once I was even called 'mean' by another member on the board and I was like 'no – I just ask questions that need to be answered. I'm sorry if my tone of voice sounds different to your tone of voice but I know that I'm not being mean.' So there is definitely this issue with women trying to be heard in the right way. – **Ms Z** The shortage of women in tech has mostly spurred me on but that is probably due to my personality type. – **Ms Y** It's just about being able to speak up louder because chances are you can do it, you're just not always looked at first to give it a shot. It comes down to self-confidence, insurance and knowledge of the industry. – **Ms W** Women tend to create environments that are less competitive and more cooperative and collaborative. They tend to be more nurturing … these are just stereotypically speaking. They are more concerned with making sure everyone is moving along in their careers and are happily progressing. It is difficult to say. I've worked with all sorts of women, some have their traits and some do not. – **Ms Z**	

Index

Note: **Bold** page numbers refer to tables and *italic* page numbers refer to figures.

For Product Safety Concerns and Information please contact our EU
representative GPSR@taylorandfrancis.com
Taylor & Francis Verlag GmbH, Kaufingerstraße 24, 80331 München, Germany

www.ingramcontent.com/pod-product-compliance
Lightning Source LLC
LaVergne TN
LVHW080656011026
849160LV00014B/1451

* 9 7 8 1 0 3 2 0 9 1 2 1 1 *